Reimagine Calculus

Simon Kim

Reimagine Calculus

For permissions, inquiries, or feedback, please contact:
reimaginecalculus@gmail.com

This book is self-published by the author.

First Edition, 2026

ISBN:978-1-0674534-1-1

For my family, Agnes, Cecilia, Joseph, and Daniel, who inspired me along every step of this journey.

Contents

Introduction

Math is all around us. Even though you don't recognize it, math concepts are dipped in our daily lives, and we live with it. You can see the other side of math when you find how the mathematical principles are connected to our life.

No matter how much I emphasized the importance and practicality of math, my students didn't enjoy it. As a math teacher, I hoped to deliver math concepts to my students more effectively. This prompted me to ponder how to make math more relatable and accessible to them.

After careful observation and reflection, I found a powerful tool to add to my teaching arsenal: real-life examples. By using simple instances of mathematical concepts that we encounter in our daily lives, I found that students became more receptive to learning math. They even started to develop an interest in it. This approach not only helps them understand calculus but also paves the way for a deeper comprehension of other mathematical principles. When I connect mathematical concepts to real-life analogies, students respond positively to math, like: 'Mathematics is useful in this manner' rather than 'Mathematics is boring and difficult.'

My motivation for writing this book is rooted in a common issue. I've observed that many people react negatively to calculus, often because they haven't learned it from a practical perspective. This perception, I believe, is a barrier that prevents many from fully appreciating the beauty and usefulness of calculus. I am passionate about changing this. I want to share the essence of differentiation and calculus with as many people as possible, showing them new aspects of calculus through relatable examples from our daily lives.

One of the most rewarding aspects of teaching is witnessing my students' unique applications of mathematical skills to solve real-world problems. These applications are not just theoretical; they have a direct connection to our lives. For instance, I remember a student who used math to explain the occurrence of a flood in their area. Another student modeled the spread of disease in our community. Yet another student explained the mathematical expressions for the various shapes of leaves and shells we see around us. These experiences reaffirm my belief in the power of math and its practicality in our lives, and I hope it reassures you, too, that math is not just a subject but a tool that can empower you in various aspects of life.

Many students feel intimidated by the word 'calculus' when they first start studying it. In this book, I aim to show how familiar and accessible calculus can be.

Part I
The Mathematical Language of Change

Chapter 1

The Language of Change

Math is a language. Languages such as English use consonants and vowels to create words and connect these words to make sentences to communicate with people. However, mathematics employs numbers, letters, and symbols to express and convey meaning through logical reasoning. Calculus is the mathematical language that represents **change**. It helps us understand and describe the shifting world around us. Change is everywhere, from the rising and falling temperatures each day to the unpredictable movement of stock prices and the rhythmic beating of our hearts. Calculus provides the tools to analyze and express these changes precisely, making it one of the most powerful branches of mathematics.

1.1 The Nature of Change

You are starting a new business venture and wish to acquire one of the following companies. The graph below illustrates the sales of both companies over the past five years. Which company would you select if both had the same acquisition price?

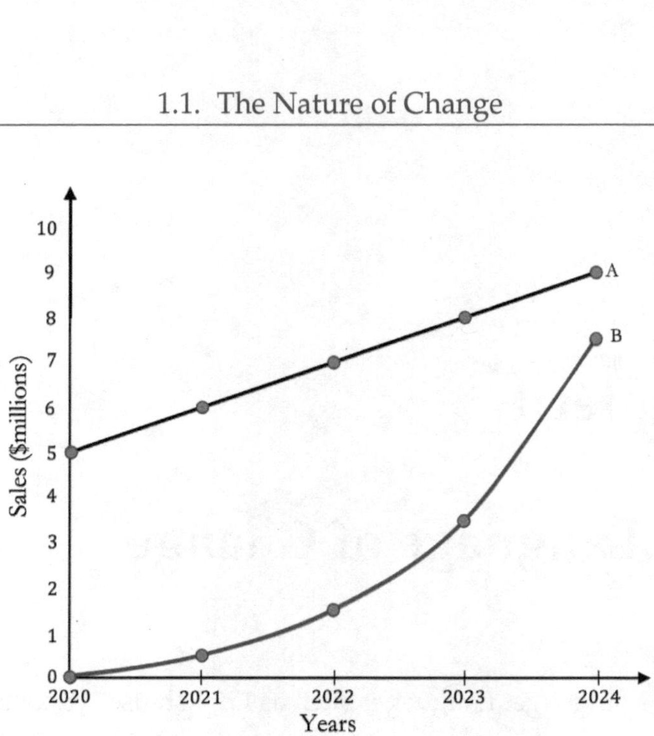

Someone interested in buying and selling the company within 2-3 years might prefer Company A, especially if they seek quick profit. Conversely, those aiming for long-term success in a rapidly expanding company may choose Company B. Company B's sales are not just steadily increasing but are experiencing exponential growth. This kind of growth is important because it means the company's value could multiply quickly over time. Over the past five years, Company A's sales have consistently increased, which is a great sign of growth. The phrase 'growing in sales' describes this kind of change. Meanwhile, Company B's sales initially grew at a slow pace but have accelerated recently, indicating an improvement compared to its past performance. Words like 'increasing slowly' and 'increasing rapidly' help us understand how sales change over time. Overall, the graph illustrates how sales have changed over the years.

1.2 Linear vs. Nonlinear Change

To understand change mathematically, you need to become familiar with the language of mathematics, which includes numbers, symbols, and letters. Let's explore how this mathematical language expresses change.

The graph shows the sales growth over time. We can define time as the input value and sales as the output value in the graph. As mentioned, math prefers letters or symbols. So, let the input value be x, the output value be y, and the change or difference be Δ(delta). The mathematical expression for the change of time will be Δx, and the expression for the change of sales will be Δy. Therefore, we can express the change in Company A's sales over time as follows:

$$\frac{\text{change in sales}}{\text{change in time}} = \frac{\text{sales at time 2 - sales at time 1}}{\text{time 2 - time 1}} = \frac{\Delta y}{\Delta x}$$

In mathematics, it is generally logical to place the **independent variable**, such as time or age, in the denominator of a fraction. In contrast, the **dependent variable**, such as sales or weight, is placed in the numerator. This convention reflects how the output (dependent variable) changes in response to the input (independent variable). Since time is the independent variable, it is placed in the denominator, while sales — which depend on time and other influencing factors — are placed in the numerator. Therefore, the change in Company A's sales over time is expressed as the **rate of change** or **slope**:

$$\frac{\Delta y}{\Delta x}$$

Let's quantify the change in sales for Company A with some numerical values. If Company A achieved sales of \$5 million in 2020 and \$6 million in 2021, the yearly sales increase would be expressed as follows:

- Sales change between 2021 and 2020

$$\frac{\Delta y}{\Delta x} = \frac{(\$6 - \$5) \text{ million}}{2021 - 2020} = \$1 \text{ million}$$

There was a one million dollar increase in sales between 2020 and 2021. If the sales were \$6 million and \$7 million in 2021 and 2022, respectively, the change of sales over time between 2022 and 2021 will be calculated as follows.

- Sales change between 2022 and 2021

$$\frac{\Delta y}{\Delta x} = \frac{(\$7 - \$6) \text{ million}}{2022 - 2021} = \$1 \text{ million}$$

The change in sales over the years from 2022 to 2024 can be summarized below.

- Sales change between 2023 and 2022

$$\frac{\Delta y}{\Delta x} = \frac{(\$8 - \$7) \text{ million}}{2023 - 2022} = \$1 \text{ million}$$

- Sales change between 2024 and 2023

$$\frac{\Delta y}{\Delta x} = \frac{(\$9 - \$8) \text{ million}}{2024 - 2023} = \$1 \text{ million}$$

The sales growth rate stays the same as before, at \$1 million per year, showing that company A's sales trend stays consistent each year. In mathematics, the 'growth trend' is expressed as a 'rate of change' or 'slope'. Therefore, we can say that Company A's rate of change or slope for sales over time is \$1 million per year, indicating that Company A's sales have remained steady for the past five years.

The sales changes in Company B are more dynamic than those in Company A. The extent of change in sales for Company B is notable. Company B's sales were 0 in 2020 and \$0.5 million in 2021, the change in sales can be expressed as follows:

- Sales change between 2021 and 2020

$$\frac{\Delta y}{\Delta x} = \frac{(\$0.5 - \$0) \text{ million}}{2021 - 2020} = \$0.5 \text{ million}$$

Company B has shown a steady increase in sales every year. It is summarized below.

- Sales change between 2022 and 2021

$$\frac{\Delta y}{\Delta x} = \frac{(\$1.5 - \$0.5) \text{ million}}{2022 - 2021} = \$1 \text{ million}$$

- Sales change between 2023 and 2022

$$\frac{\Delta y}{\Delta x} = \frac{(\$3.5 - \$1.5) \text{ million}}{2023 - 2022} = \$2 \text{ million}$$

- Sales change between 2024 and 2023

$$\frac{\Delta y}{\Delta x} = \frac{(\$7.5 - \$3.5) \text{ million}}{2024 - 2023} = \$4 \text{ million}$$

Therefore, we can conclude that Company B's sales growth rate has doubled each year over the past five years. "How can we project the sales of both companies in 2025?" Based on the data in the table below Company A has maintained a steady, linear growth pattern, which predicts sales of about $10 million in 2025. In contrast, Company B's growth rate has doubled annually since 2021, leading to a projected sales figure of approximately $15.5 million in 2025.

Company A

Year	Sales
2020	$5 million
2021	$6 million
2022	$7 million
2023	$8 million
2024	$9 million
2025	?

Slope: +$1 million, +$1 million, +$1 million, +$1 million, +$1 million

Company B

Year	Sales
2020	$0 million
2021	$0.5 million
2022	$1.5 million
2023	$3.5 million
2024	$7.5 million
2025	?

Slope: +$0.5 million, +$1 million, +$2 million, +$4 million, +$8 million

The tables show that Company B's slope varies each year, while Company A's slope stays constant. Mathematically, the sales of Company A are shown by a straight line with a fixed slope. In contrast, Company B's sales are represented by a curve where the slope changes over time.

The slope of Company A's sales over the past five years can be described as follows:

$$\frac{\Delta y}{\Delta x} = 1$$

Meanwhile, the slope of Company B's sales over the last five years cannot be represented by a single constant value:

$$\frac{\Delta y}{\Delta x} \neq 1$$

As discussed earlier, a line and a curve represent different types of change. In mathematics, **differentiation**, a key concept in calculus, is a fundamental tool for analyzing how these changes occur by identifying patterns in their slopes. However, before learning and studying the complex patterns associated with differentiation, we must understand **functions**, which are the fundamental building blocks of differentiation in calculus.

1.3 Functions: The Language of Change

We frequently encounter situations that can be mathematically modelled using functions. Some relationships follow a linear pattern, such as the sales performance of Company A in the previous section or an object moving at a constant speed. However, many real-world scenarios require nonlinear functions, including Company B's fluctuating sales or the movement of a car that accelerates and decelerates.

To mathematically represent change patterns, we first need to understand functions, which are the foundation of calculus. Once we understand functions, we can move on to studying differentiation. A **function** defines a precise relationship between inputs and outputs, ensuring that **each input corresponds to exactly one output**. One way to understand this is by using a **function machine**, a conceptual tool that processes input according to a specific rule. When the same raw materials are fed into the system, the machine applies the same steps each time, producing consistent results. Similarly, in mathematics, a function consistently applies its rule to every input.

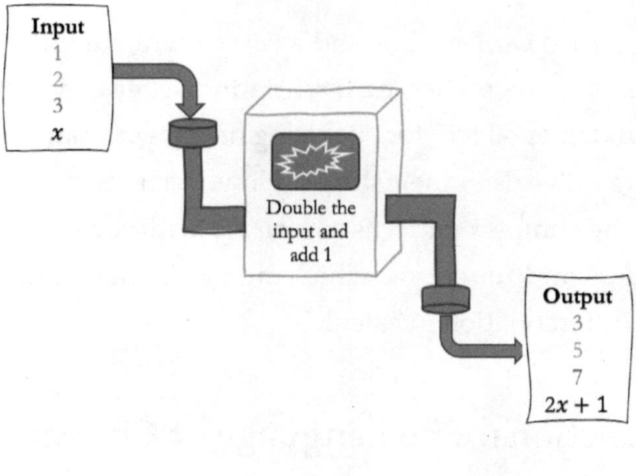

The function machine introduced above, which doubles the input and then adds 1 to compute the output, is commonly represented by f in mathematical notation. In this diagram, the box itself represents the function — the rule that converts inputs into outputs. Functions like this form the foundation of calculus and allow us to describe and analyze change systematically.

Once again, mathematics is a language. Just as we can translate ideas from English into another spoken language, it is possible to translate English into a mathematical language. For example, the sentence 'The function above doubles the input value and then adds 1' can be mathematically expressed as $f : x \to 2x + 1$. In this expression, every letter, number, and symbol has a precise meaning and is logically connected to the others, as illustrated below.

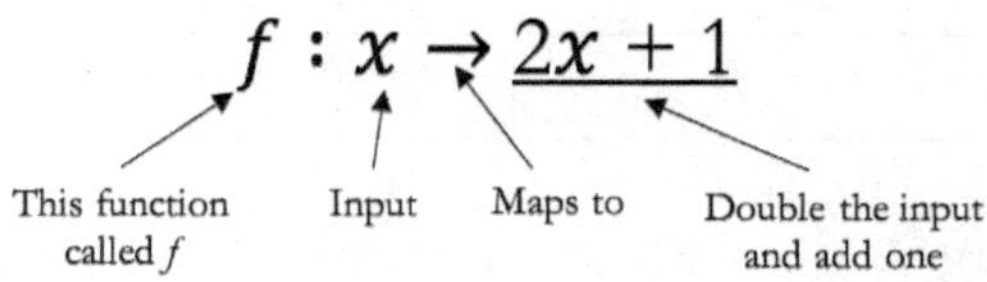

Hence, we can understand mathematics more deeply by becoming familiar with the language used to express functions. For example, the function written as "$f : x \to 2x + 1$" can also be written as "$f(x) = 2x + 1$." Both notations are equivalent, but in this book we will mainly use the second form, since it is more commonly used and easier to work with.

1.4 Linear Function vs. Nonlinear Function

In the function, x and $f(x)$ represent the input and output values, respectively. The output values vary depending on the input values. For this reason, the input values are called 'independent variables', and the output values are called 'dependent variables'.

The function, $f(x) = 2x + 1$, can also be expressed using a table shown below. When representing a function in a table, the input values should appear on the left side, while the output values should be shown on the right side.

Input x	Output $f(x) = 2x + 1$
1	$2(1) + 1 = 3$
2	$2(2) + 1 = 5$
3	$2(3) + 1 = 7$
a	$2(a) + 1 = 2a + 1$
x	$2(x) + 1 = 2x + 1$

The table above displays that the function is a **linear function** due to its constant rate of change. The output difference remains consistent with any equal change in the input.

In other words, the function maintains a constant change between input and output values. This relationship can be visualized using the table below, where the differences between consecutive outputs remain unchanged.

Δx	Input x	Output $f(x) = 2x + 1$	Δy
	1	3	
+1	2	5	+2
+1	3	7	+2
+1	4	9	+2
+1	5	11	+2

Additionally, when we plot $f(x)$ on the xy-plane, the function forms a straight line — another defining characteristic of linear functions, as the graph appears as a straight line.

A straight line is a type of graph where the "steepness," or rate of change, remains constant at every point. This characteristic makes linear functions easy to analyze. Later, as we examine curves, we will see that their slopes differ from point to point, which is why calculus is necessary to measure those changing rates of change.

In the linear function $f(x) = 2x + 1$, the change between output values for successive input values is always two since the rate of change $\frac{\Delta y}{\Delta x}$ is constant. This means the function's slope is constant, as shown in the graph below.

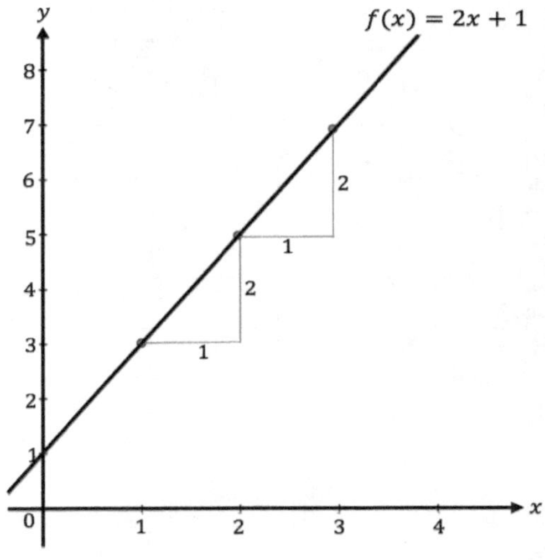

Typically, mathematical functions, represented by $f(x)$, are expressed as y on a graph. These two expressions are interchangeable and convey the same meaning.

The straight line is used to express change at a constant rate. For example, if you get paid \$15 per hour, you can expect \$30 for two hours or $150 for 10 hours since your job's income is based on the fixed hourly rate you work. However, not all functions behave this way. Many real-world phenomena, such as acceleration, population growth, and the motion of objects under gravity, are best described using nonlinear functions, in which the rate of change varies at different points.

A simple example of a **non-linear function** is shown in $f(x) = x^2$. Unlike a linear function, where the change in output remains consistent for equal changes in input, the function $f(x) = x^2$ has an inconsistent slope — meaning the rate of change of the function's output is not constant but continuously increasing, as shown below in the table and graph.

Δx	Input x	Output $f(x) = x^2$	Δy
	1	1	
+1	2	4	+3
+1	3	9	+5
+1	4	16	+7
+1	5	25	+9

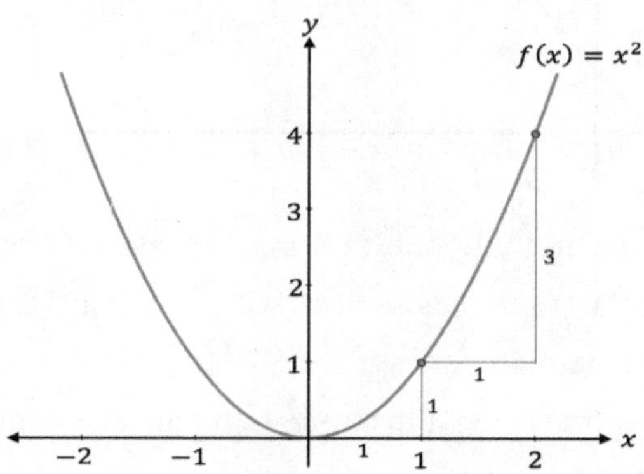

To understand this more clearly, look at how the output values change as x increases. The changes in y-values (Δy) are not constant, but the changes of those changes — written as $(\Delta(\Delta y))$ — do follow a pattern. For example, in the function $f(x) = x^2$, the value of $(\Delta(\Delta y))$ is always 2. This shows that even though the rate of change is not constant, it still follows a predictable structure that we can analyze mathematically.

Δx	Input x	Output $f(x) = x^2$	Δy	Δ of Δy
+1	1	1	+3	+2
+1	2	4	+5	+2
+1	3	9	+7	+2
+1	4	16	+9	
	5	25		

Recognizing these patterns is crucial because it enables us to describe how curves behave and define a precise slope for a curve at any point. This idea forms the foundation for understanding **derivatives**, which offer a way to measure instantaneous change — one of the core concepts in calculus.

Understanding the slope of a curve is not merely a mathematical exercise; it has practical applications in predicting and preparing for real-world changes. Because nonlinear functions exhibit a changing rate of change, analyzing their slopes enables us to anticipate future trends.

For example, in a classroom setting, tracking the rate of students catching a cold over time can help predict future cases. On a larger scale, understanding how the number of infections changes during a pandemic can assist in planning resources and containment strategies. Similarly, businesses can benefit from analyzing trends in irregular product sales, which helps them optimize inventory and reduce costs. In urban planning, forecasting population growth can lead to better infrastructure planning and more efficient budget allocation.

These examples emphasize the significance of predicting the rate of change in non-linear functions. Recognizing patterns in change can assist us in making informed decisions, solving problems more effectively, and preparing better for the future. This ability to analyze and quantify change is a core component of calculus, paving the way for a deeper exploration of instantaneous rates of change.

1.5 Checkpoint 1

1. Adam called a plumber to fix a plumbing issue at his home. The plumber informed him there was a \$100 travel fee and an additional \$60 per hour charge.

   (a) Adam is trying to determine Δx and Δy to find the average rate of change of cost over time in the plumber's work. What would Δx and Δy represent?

   (b) How can he determine the rate of change in cost over time?

   (c) Adam developed a formula to calculate the payment for a plumber's three hours service. How would he create this equation?

2. Adam needed to pay the plumber \$100 for the initial fee and an additional \$60 for each hour of work, and he tried to express these facts using a function machine.

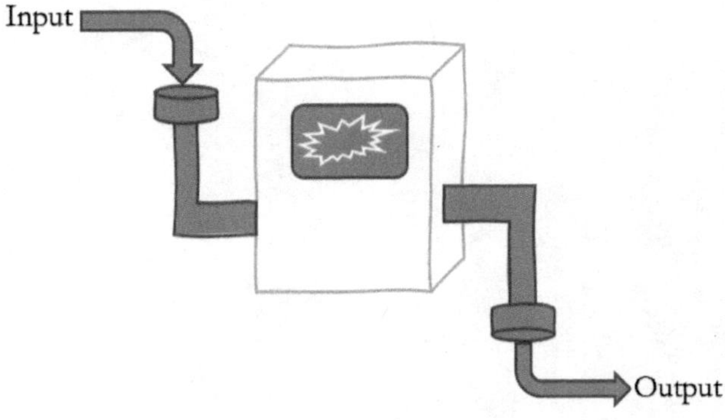

   (a) What would be the input value of the function machine?

   (b) What expression should be in the function machine?

(c) What would the output of the function machine be when the input is 1, 2, and 3?

3. Adam makes $f(x) = 60x + 100$ to express the situation above.

    (a) What do $f(x)$ and x represent in the context of the problem?

    (b) Can you explain what the numbers 60 and 100 mean in relation to the situation in the above function?

4. Adam believes that the payments change constantly, and can be represented as a straight line. Is he correct? Justify your answer.

1.6 Checkpoint 1 solution

1. (a) In this context, Δx represents the change in time (in hours) that the plumber spends working, and Δy represents the change in the total cost (in dollars) of the plumbing service.

   (b) The rate of change is expressed as $\frac{\Delta y}{\Delta x}$

   $$\text{Slope} = \frac{\Delta y}{\Delta x} = \frac{\$60}{\text{hour}}$$

   (c) The cost over time can be calculated using the following expression:

   $$\text{Payment for service hours} = 60\frac{\$}{hour} \times (\text{hour}) + 100$$

   The above expression can be converted to a function as follows:

   $$f(x) = 60x + 100$$

   where x represents the number of service hours. For three hours of service:

   $$f(3) = 60(3) + 100 = 280$$

   The total payment for three hours of service is \$280

2. (a) The number of hours is the function's input value, as it is an independent variable.

   (b) The expression × 60 and + 100 would be in the function machine since the hourly additional cost is \$60, and the initial cost is \$100.

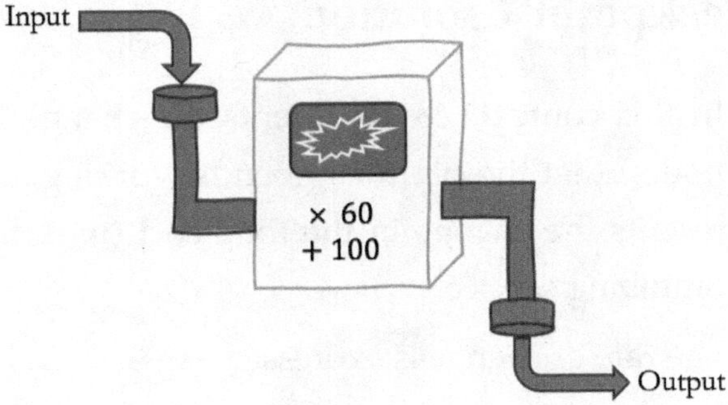

(c) When the input values 1, 2, and 3 are entered into the function machine, the output values will be 160, 220, and 280, respectively.

3. (a) $f(x)$ represents output as the payment, while x represents the time worked in hours.

   (b) The number 60 represents the rate of change, or the hourly rate, which is the slope. The number 100 represents the initial payment. This initial cost does not change over time since it is a one-time cost at the beginning that is the same for any length of service.

4. The payment increases at a constant rate of 60 over time, so the payment over time is a linear function. A nonlinear function would show curve behavior, continuously changing the slope with the input values.

Chapter 2

The Slope of a Tangent Line

In the previous chapter, we introduced instances of nonlinear situations, often encountered in the real world, such as the growth rate of an infectious disease. This chapter will discuss the mathematical techniques for determining the slope of nonlinear functions. We'll introduce the concept of limits and explore the geometric idea of the tangent line, providing a foundation for understanding how to find the slope of a curve at any given point.

2.1 Understanding Slope on a Curve: Two Approaches

In geometry, a line that intersects two distinct points is called a **secant line**, while a line that touches a curve at a point is called a **tangent line**.

There are two ways to express the slope of a curve, as shown in the diagram below. The first method involves connecting two points on the curve and calculating the change, referred to as the

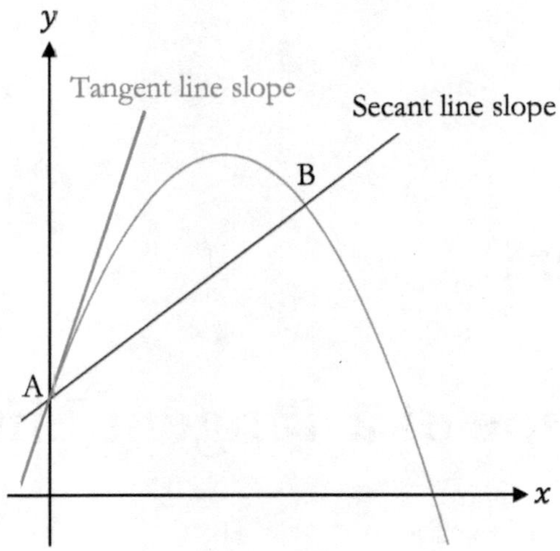

slope of the secant line. The second method entails finding the slope at a single point on the curve, known as the slope of the tangent line.

Finding the slope between two points is easier than finding it at a single point. Let's start by discussing how to find the slope between two points. Afterward, we will explore how to determine the slope of a curve at one point.

The formula for the slope of the secant line on a curve is equivalent to the rate of change or slope between two points.

$$\text{Rate of change between two points} = \frac{\Delta y}{\Delta x}$$

The difference between x and y values in a curve is represented as Δx and Δy, respectively. The graph below shows that. Δx is the difference in the x-axis, $b - a$, and the Δy is the difference in the y-axis, $f(b) - f(a)$.

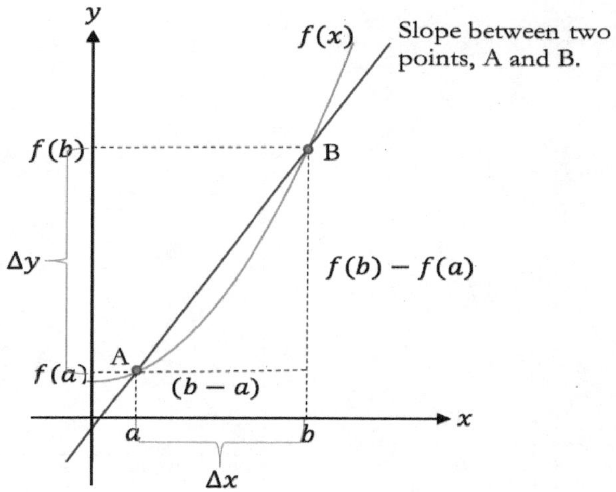

The **slope of the secant line** on the curve will be

$$\text{The slope between two points} = \frac{\Delta y}{\Delta x} = \frac{f(b) - f(a)}{b - a}$$

This equation is also called the **average rate of change** between two coordinates, $A(a, f(a))$ and $B(b, f(b))$, since it represents the average or overall change over the specified interval from a to b.

2.2 Limit: The Key to the Tangent Line

The **tangent slope** refers to the slope at a single point on a function, where the tangent line just touches the curve and shows how steeply it is changing at that exact spot.The graphs on the left show the slope between two distinct points on a nonlinear function $f(x)$.

The graph on the right shows how the slope of the secant line changes as b gets closer to a on the x-axis. As b moves toward a, the slopes of the secant lines between points A and B_1, B_2, B_3, and

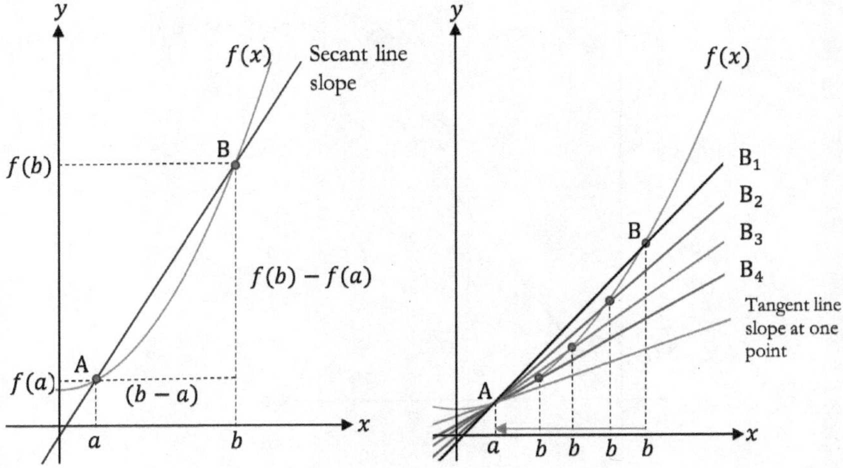

B_4 also change. The slope of the secant line between A and B_4 is much closer to the value of the slope of the tangent line at A than the slope of the secant line between A and B_1. This observation leads to an important question: What happens to the slope of the secant line as b gets infinitely close to a? The answer is:

"As b moves infinitely close to a, the secant line slope gets closer and closer to the tangent line slope at A. Eventually, they become the same."

This idea is known as a **limit**, occurring when one point approaches infinitely close to another. Understanding the meaning of 'approaching infinitely' seems complicated, but it's not as difficult as you might think. A simple example of getting closer to infinity is when 0.9 approaches 1, as shown below.

$$0.9 \rightarrow 0.99 \rightarrow 0.999 \rightarrow 0.9999 \rightarrow 0.99999 \rightarrow \cdots 1$$

This indicates that one value can change continuously and approach another infinitely closely. This naturally leads to an important question:

"If one number becomes infinitely close to another, are the two numbers equal?"

What do you think? If a point moves progressively closer to another, indefinitely reducing the distance between them, do they eventually become the same point? In math, the answer depends on how we define "closeness." A number can get infinitely close to another without ever actually reaching it. No matter how small the gap is, there is always a tiny distance between them — unless we say they are equal. This idea is fundamental in calculus. It helps us understand limits, which describe how numbers or functions behave as they get closer to a certain value.

Let's prove that 0.99999... and 1 are equal through a simple deduction. Consider the fraction $A = \frac{1}{3}$ and its decimal represented $B = 0.33333\ldots$ in the table below. Since these are two ways of expressing the same value, we know that $A = B$.

Claim 1	$A \times 3 = 1$	$\frac{1}{3} \times 3 = 1$
Claim 2	$A = B$	$\frac{1}{3} = 0.33333\ldots$
Claim 3	$B \times 3 = 1$	$0.99999\ldots = 1$

In Claim 1, A multiplied by 3 equals 1 since $\frac{1}{3} \cdot 3$ is 1. In Claim 2, we state that A and B are equal, and we know this is true. Therefore, B multiplied by 3 should be equal to 1. Consequently, 0.33333... multiplied by 3 equals 1, which means that $0.99999\ldots$ equals 1. Although a more formal proof would require advanced concepts like sequences and series, this simple argument helps illustrate that numbers can have different representations but still be equal.

We demonstrated that as one value approaches infinitely, it becomes equal to another, as shown as $0.99999\ldots$ becomes 1. For the same reason, the secant line can become the tangent line on a curve as shown in the graph below. A point b on the x-axis approaches infinitely close to point a, the secant lines between points A and B, indicated by B_1, B_2, until B_∞, approach infinitely close to the tangent line at point A. The B_∞ represents the secant line infinitely close to the tangent line. The slope of the secant line for B_∞ is equal to that of the tangent line.

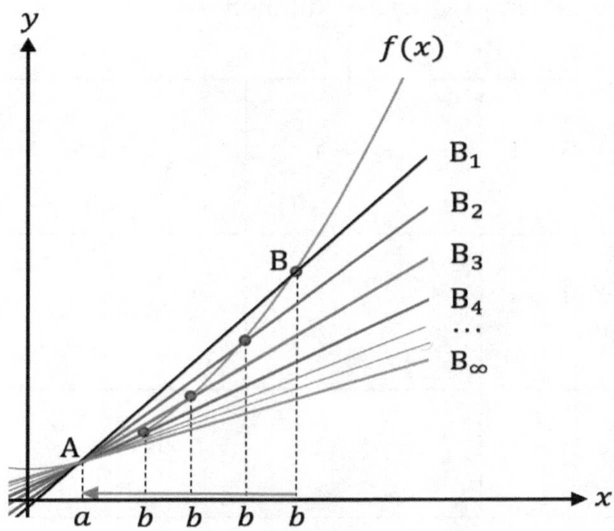

In conclusion, finding the slope at a single point begins with the slope of the secant line. We need to bring the two points closer together to obtain a more precise value. This is where limits come into play. By using limits, we transition from an average rate of change to a more exact rate at a specific point. This concept is a key stepping stone in understanding how slopes function in calculus. Therefore, we should learn how to connect these two concepts mathematically.

2.3 Exploring the Limit of a Function

As we discussed, finding the slope at a single point requires understanding limits and how they interact with functions. In this section, we will explore the relationship between functions and limits, gaining a deeper understanding of their meaning and significance.

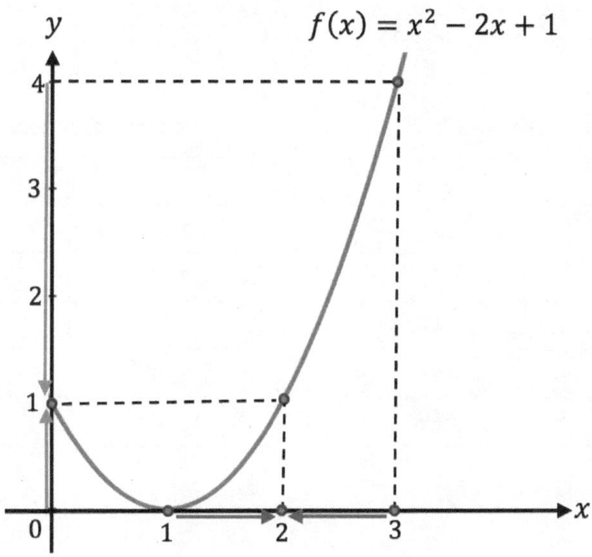

A nonlinear function $f(x) = x^2 - 2x + 1$ is shown above. We investigate the behavior of $f(x)$ as x approaches 2 using a graph and a table. The graph above illustrates how the limit operates visually. The x values between 1 and 2 approach 2 from the left, causing the $f(x)$ value to increase from 0 to 1. The x values between 3 and 2 approach 2 from the right, resulting in the function $f(x)$ decreasing from 4 to 1. Thus, we can conclude that the function $f(x)$ approaches 1 as x nears 2 from both sides.

The table below shows a different way that the value of $f(x)$ approaches 1 as the value of x approaches 2 from the left or the right side. The values of $f(x)$ get closer to 1 as the values of x approach 2 from the left. Conversely, the table on the right demonstrates how the results of $f(x)$ get closer to 1 when x approaches 2 from the right side, particularly in the transition from 3 to 2. Based on the values provided in the table, we can conclude that the function $f(x) = x^2 - 2x + 1$ converges to 1 as x approaches 2 from either side.

x approaches from 1 to 2

x	$f(x)$
1.0	0
1.5	0.25
1.9	0.81
1.99	0.98
1.999	0.998

x approaches from 3 to 2

x	$f(x)$
3.0	4
2.5	2.25
2.1	1.21
2.01	1.02
2.001	1.002

We can see from the graph and the table above that the value of $f(x)$ approaches 1 as x approaches from either the left or the right side of 2. This situation is expressed as follows:

$$\lim_{x \to 2} f(x) = 1$$

In other words, when the function values of $f(x)$ approach a certain value of L as x nears a, this can be expressed as follows:

$$\lim_{x \to a} f(x) = L$$

It reads, **the limit of $f(x)$ is equal to L, as x approaches a,** or **$f(x)$ approaches L as x approaches a.**

Let's take a closer look at the language of calculus presented above. Using the table below, we can translate English into the language of mathematics and vice versa.

English	Mathematical language
function name	f
input value	x
close or approaches	$\rightarrow$
x infinitely close to a	$\lim_{x \to a}$
the same or equal	$=$
output value at $x = a$	$f(a)$ or L

The following shows how the statements translate to mathematical language:

- Statement: "A function named f with an input value x."
  Mathematical expression: $f(x)$

- Statement: "The input value x infinitely approaches a"
  Mathematical expression: $\lim_{x \to a}$

- Statement: "The output value is $f(a)$ or L"
  Mathematical expression: $f(a)$ or L

Combining these ideas, we can say: "As the input value x gets infinitely close to a, the output value $f(x)$ approaches L, which may be equal to $f(a)$." The mathematical expression for this is:

$$\lim_{x \to a} f(x) = L$$

or

$$\lim_{x \to a} f(x) = f(a)$$

The equations introduced above are a prime example of how mathematical language can clearly express complex situations using letters or symbols. While English conveniently allows for detailed explanations to clarify concepts, mathematics relies on a logical structure to convey meaning efficiently. For this reason, logic in mathematical expression is the key element that connects symbols and letters. Understanding the meaning of these symbols and translating them into English can make learning mathematics much more enjoyable. With this foundation, we are now ready to connect these ideas back to the tangent line slope, using our knowledge of functions and notation to describe change more precisely.

2.4 Using Limits to Find the Slope at a Single Point

To understand how to determine the slope at a single point, we begin by recalling the two-point slope formula. Given two points on a function, the slope of the secant line passing through these points is defined by:

$$\frac{\Delta y}{\Delta x} = \frac{f(b) - f(a)}{b - a}$$

However, we encounter a challenge when attempting to calculate the slope at a single point. The setup in the formula leads to:

$$\frac{f(b) - f(a)}{b - a} = \frac{f(a) - f(a)}{a - a} = \frac{0}{0}$$

This is an **indeterminate value**, meaning it cannot be determined, or a meaningful solution cannot be found. This problem arises because the secant formula requires two different points. If we use the same point twice, we end up dividing by zero.

To illustrate this, consider the function $f(x) = x^2$ at the point $x = 1$, which corresponds to the point $(1, 1)$.

$$\frac{f(1) - f(1)}{1 - 1} = \frac{1 - 1}{1 - 1} = \frac{0}{0}.$$

This result is undefined, $\frac{0}{0}$, and is called an **indeterminate form**. For example, $0 \cdot \infty$ or $\frac{0}{0}$ are indeterminate forms since both are the same. The expression $\frac{0}{0}$ can be separated as $0 \cdot \frac{1}{0}$ and is equal to $0 \cdot \infty$. The two values of 0 and ∞ are like a tug-of-war at opposite ends, so we cannot determine the exact value. However, as shown in the graph below, the slope of the tangent line, which is a slope at a single point, exists at $x = 1$. To address indeterminate forms, we use algebraic techniques, such as simplifying expressions by cancelling common factors. Let's look at an example to understand this process.

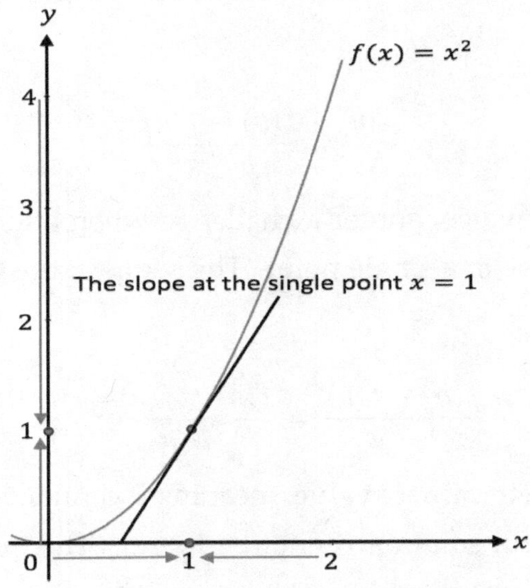

The function we will use is $f(x) = x^2$, which is shown in the above graph. By applying the concept of a limit, we can examine how values behave as they approach a specific point, such as $a = 1$. Here, we set $a = 1$ as our reference point and let x be a variable approaching 1.

We summarize this idea as follows:

$$x = x, f(x) = x^2$$

$$a = 1, f(1) = 1$$

These values can be substituted into the formula for the slope of the secant line between these two points, then we will have the following:

$$\frac{f(x) - f(a)}{x - a} = \frac{x^2 - 1}{x - 1}$$

If x equals 1 in the formula above, the result is an indeterminate value of $\frac{0}{0}$. Therefore, the expression is undefined at $x = 1$, making it a non-permissible value. As a result, we cannot cancel the common factor $x - 1$. However, if we applied the limit, the result would be different.

When we take the limit as x approaches 1, and after recalling the above and applying the limit in the formula, we obtain the following:

$$\lim_{x \to 1} \frac{f(x) - f(1)}{x - 1} = \lim_{x \to 1} \frac{x^2 - 1}{x - 1}$$

Now, we can simplify the expression by factoring and taking the limit, since the limit considers x approaching 1 but not equal to 1. Therefore, after factoring, we can cancel the common factor and evaluate the limit. This calculation successfully finds the slope at the point $x = 1$ as 2 using limits.

$$\begin{aligned} \lim_{x \to 1} \frac{x^2 - 1}{x - 1} &= \lim_{x \to 1} \frac{(x+1)\cancel{(x-1)}}{\cancel{x-1}} \\ &= \lim_{x \to 1} (x+1) \\ &= 2 \end{aligned}$$

In summary, by combining the concept of limits with the average rate of change, we arrive at the formula for the slope of the tangent line:

$$\lim_{x \to a} \frac{f(x) - f(a)}{x - a}$$

It represents that as x approaches a, the secant line slope $\frac{f(x)-f(a)}{x-a}$ approaches the tangent line slope at $x = a$.

The graph below visually represents the mathematical expression. As x approaches a, the point Q moves along the curve, getting closer to the point P. The secant line that passes through these points approaches the tangent line at P.

Thus, the formula, $\frac{f(x)-f(a)}{x-a}$ as x approaches a, represents the slope at a specific point P on the curve, making it the **instantaneous rate of change** as it arises when the average rate of change tends to an instant. The instantaneous rate of change is a key concept in calculus. This idea will be examined in more detail in Chapter 4.

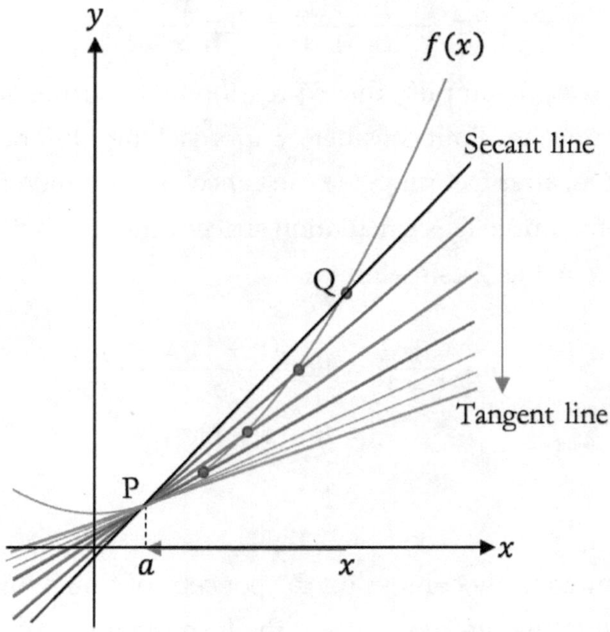

2.5 Checkpoint 2

1. The graph below illustrates a function $f(x)$ with a point $P(0,2)$ and various other points $Q = (x, f(x))$ along the curve. The slope of the tangent line at point P is 2. Using the coordinates of point Q, complete the table below to determine the slope of the secant line between P and Q. This will help you observe how the slope of the secant line approaches that of the tangent line slope — the instantaneous rate of change.

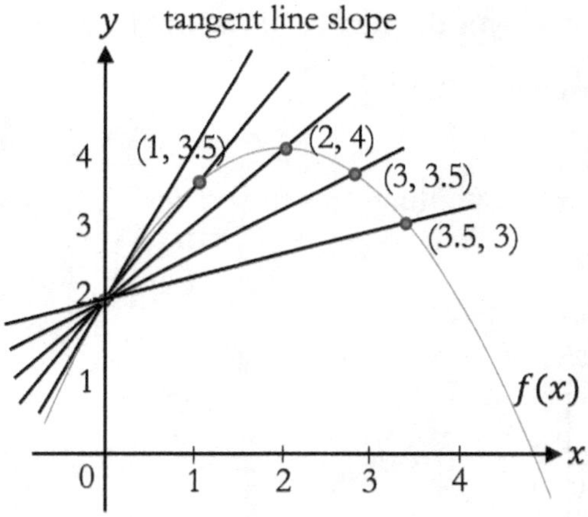

Coordinate P	Coordinate Q	Δx	Δy	$\frac{\Delta y}{\Delta x}$
(0, 2)	(3.5, 3)	3.5	1	$\frac{1}{3.5} = 0.29$
(0, 2)	(3, 3.5)			
(0, 2)	(2, 4)			
(0, 2)	(1, 3.5)			

2. Adam, a biologist, is busy cultivating bacteria for his lab experiment. He discovered that just $1mm$ of bacterial growth in the medium corresponds to about one million bacteria. To study how the colony grows over time, Adam uses the following function, where $S(t)$ (in millimeters) represents the size of the colony after t hours:

$$S(t) = 0.1t^2$$

$S(t)$: the size of the bacterial colony in the medium, in mm

t : time, in hours

The graph below illustrates the bacteria's growth over the last few hours.

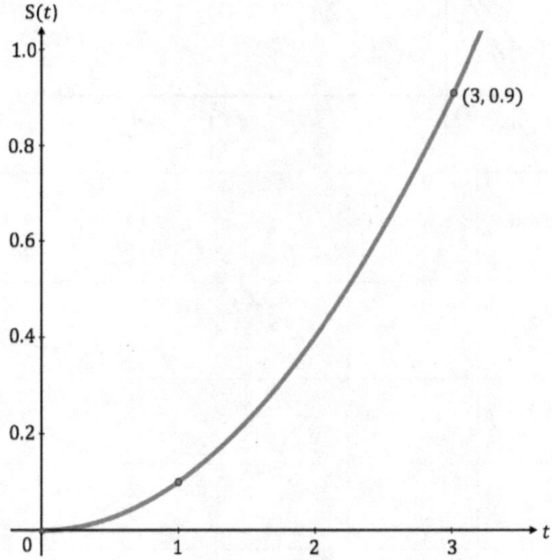

(a) How can Adam calculate the average rate of change in the bacterial population over the time interval from 0 to 3 hours?

    i. What does this average rate tell us about the overall growth during that time?

(b) How can he express this average rate of change as a secant line on the graph?

    i. What does the slope of this secant line signify in the context of the bacteria's growth?

(c) How can Adam determine the instantaneous rate of change at exactly 1 hour?

    i. Why might this differ from the average rate over the time interval?

(d) How is the instantaneous rate of change represented as a tangent line on the graph?

    i. What does the slope of the tangent line represent in this context?

(e) Why is the instantaneous rate of change beneficial to Adam as a biologist?

    i. How can it help him make predictions or decisions for his experiment?

3. Recall the expression $\frac{x^2-1}{x-1}$ in Chapter 2.4. Why can't we simplify the common factor $x-1$ in the expression $\frac{x^2-1}{x-1}$ without limit?

2.6 Checkpoint 2 Solution

1. As the x-coordinate approaches $x = 0$, the slope gets closer to 2. If the x-coordinate becomes infinitely close, and Δx approaches zero, the slope of the secant $\frac{\Delta x}{\Delta y}$ will approximate 2.

Coordinate P	Coordinate Q	Δx	Δy	$\frac{\Delta y}{\Delta x}$
(0, 2)	(3.5, 3)	3.5	1	$\frac{1}{3.5} = 0.29$
(0, 2)	(3, 3.5)	3	1.5	$\frac{1.5}{3} = 0.5$
(0, 2)	(2, 4)	2	2	$\frac{2}{2} = 1$
(0, 2)	(1, 3.5)	1	1.5	$\frac{1.5}{1} = 1.5$

2. (a) To calculate the average rate of change over the interval, use the formula:

$$\frac{f(b) - f(a)}{b - a}$$

In our case, $f(b)$ is $S(3) = 0.1(0.3)^2$ and $f(a)$ is $S(0) = 0.1(0)^2$. Additionally, a and b are 3 and 0, respectively, as they represent the time intervals. We can calculate the average rate of change for the given time interval as follows:

$$\frac{S(3) - S(0)}{3 - 0} = \frac{0.1(3)^2 - 0.1(0)^2}{3} = \frac{0.9 - 0}{3} = 0.3\frac{mm}{hr}$$

i. The average rate of change is 0.3 mm per hour. This means that over the first 3 hours, the population increased, on average, by 300,000 bacteria per hour since 1 mm of growth in the medium represents 1 million population.

(b) The secant line shown above connects the two points on the graph:

$$\text{At } t = 0, \; S(t) = 0$$

$$\text{At } t = 3, \; S(t) = 0.9$$

i. The secant line between $t = 0$ and $t = 3$ has a slope of 0.3 $\frac{mm}{hr}$. It represents the average growth rate (bacteria population growth rate) over that three-hour interval.

(c) To find the instantaneous rate of change, we recall the formula for the slope at a single point as follows:

$$\lim_{x \to a} \frac{f(x) - f(a)}{x - a}$$

In our case, the function $f(x)$ can be replaced by $S(t)$, representing the size of the bacterial colony at time t. The variable x in the general definition can simply be replaced with another variable, such as t. Additionally, $f(a)$ can be replaced with $S(1)$ as the time approaches 1 hour. Now, we can rewrite the formula as follows:

$$\lim_{t \to 1} \frac{S(t) - S(1)}{t - 1}$$

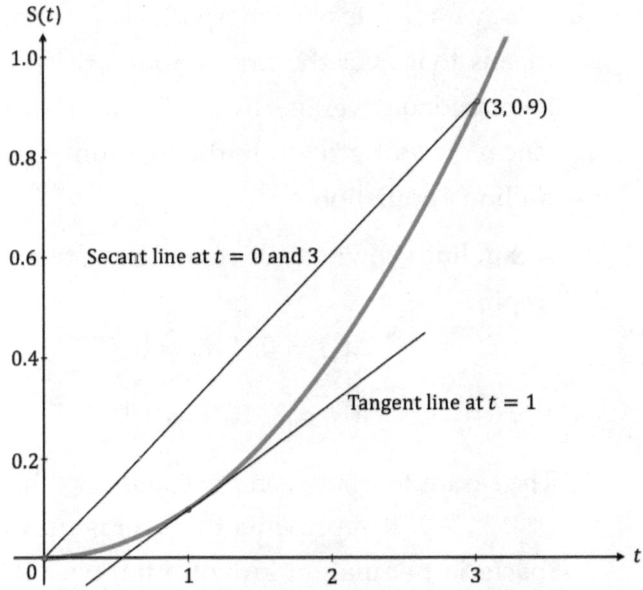

Using the function, $S(t) = 0.1t^2$, we can calculate the instantaneous rate of change as follows:

$$\begin{aligned}
\lim_{t\to 1} \frac{S(t) - S(1)}{t-1} &= \lim_{t\to 1} \frac{0.1t^2 - 0.1(1)^2}{t-1} \\
&= \lim_{t\to 1} \frac{(0.1t^2 - 0.1)}{t-1} \\
&= \lim_{t\to 1} \frac{0.1(t^2 - 1)}{t-1} \\
&= \lim_{t\to 1} \frac{0.1(t+1)\cancel{(t-1)}}{\cancel{t-1}} \\
&= \lim_{t\to 1} 0.1(t+1) \\
&= 0.2
\end{aligned}$$

The instantaneous rate of change at $t = 1$ is $0.2\ \frac{mm}{hr}$.

i. The instantaneous rate indicates the growth at exactly one moment from the above calculation. The population growth will not be constant; rather, it will be accelerated. Early growth is slower than later growth, but it increases faster over time. Therefore, Adam can accurately estimate the growth rate using the instantaneous rate of change as 0.2 $\frac{mm}{hr}$ at $t = 1$.

(d) As shown above, the tangent line intersects the curve at just one point when $t = 1$. Thus, it represents the slope at that single point, which is equivalent to the instantaneous rate of change.

i. This represents the instantaneous growth rate of the bacteria colony at 1 hour, indicating how quickly the size increases at that moment.

(e) The instantaneous rate of change allows Adam to understand the current growth rate at a specific moment instead of just an average.

i. The instantaneous rate of change enables him to predict short-term population behavior, adjust conditions for optimal growth, and make timely decisions during experiments.

3. The expression $\frac{x^2-1}{x-1}$ can be factored algebraically as $\frac{(x-1)(x+1)}{x-1}$. There is a very important condition for the $x \neq 1$, since $x = 1$ is a non-permissible value. If $x = 1$, then we are dividing by zero, $\frac{0}{0}$, which is an indeterminate value. Therefore, the common term can be cancelled out only when $x \neq 1$. However, the limit allows us to cancel out the common term with $x - 1$ in the expression, since $\lim_{x \to 1}$ describes x approaching

infinitely close to 1 but not equal to 1. So, we can safely cancel out the common factor $x - 1$ with the limit.

$$\lim_{x \to 1} \frac{(x+1)\cancel{(x-1)}}{\cancel{x-1}}$$

Chapter 3

Limits and Continuity

In the previous chapters, we introduced the concept of the tangent line slope. However, before exploring more about the slope at a single point, it is important to revisit and expand on a foundational idea that makes the instantaneous rate of change possible: the **limit**. Since we briefly discussed limits earlier, we will now examine them more thoroughly to understand their essential role in calculus.

The limit describes how a function behaves as its input approaches a specific value. Since the derivative is defined based on the limit, understanding limits is essential for defining and calculating derivatives. However, limits are not always straightforward. Some functions behave unpredictably near certain points, making it impossible to evaluate their limits. This leads us to the concept of continuity.

When a function has breaks, jumps, or holes, its limit may not be well defined, making it impossible to determine the instantaneous rate of change at those points. Thus, continuity serves as a fundamental requirement for the instantaneous rate of change, one of the most important concepts in calculus. In this chapter, we will explore the properties of limits and continuous functions, explaining why continuity plays a crucial role in our further study.

3.1 Understanding Limits and Continuity

When playing music, the piano can produce the notes Mi and Fa (E and F), but it cannot create sounds in between the two. In contrast, the violin can generate any sound between these two notes. Similarly, when drawing a sketch, a continuous line is created without lifting the pen, while a discontinuous line requires lifting the pen and starting in a different spot. These analogies help us understand the concepts of continuity and discontinuity in functions.

The graphs below appear similar but have some differences. The graph on the right, $g(x)$, is not connected between points, while the points on the left graph, $f(x)$, are connected. Let's compare the limit values between unconnected and connected points.

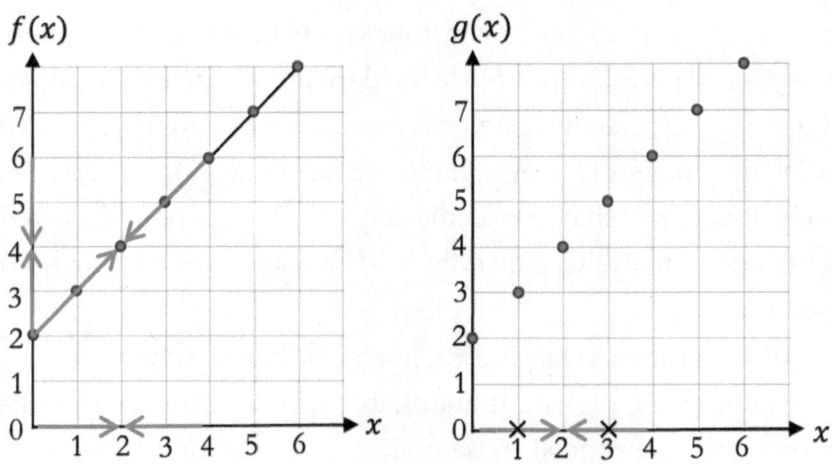

The graph $g(x)$ represents a discrete function, so the value of x cannot approach 2 from 1 or 3 because the points on the graph are separate. In the graph $g(x)$, since one value cannot lead to another through the x-axis, applying a limit is inappropriate since a function's limit means one value infinitely approaches another value. For the function $g(x)$, the **limit does not exist** (***DNE***) at $x = 2$ because the function is discontinuous and is expressed as follows:

$$\lim_{x \to 2} g(x) = DNE$$

The function $f(x)$ represents a continuous function where all the points are connected. As x approaches 2 from the left side, the value of $f(x)$ approaches 4. As x approaches 2 from the right side, the value of $f(x)$ also approaches 4. See the mathematical expression here:

$$x \text{ approaches 2 from the left side: } \lim_{x \to 2^-} f(x) = 4$$

$$x \text{ approaches 2 from the right side: } \lim_{x \to 2^+} f(x) = 4$$

The $\lim_{x \to 2^-} f(x) = 4$ indicates that the limit of $f(x)$ is 4 as x approaches 2 from the left, while the $\lim_{x \to 2^+} f(x)$ shows that the limit of $f(x)$ is 4 as x approaches 2 from the right. As a result, we can say that the **two-sided limit** approaching $x = 2$ shows the same as $f(x) = 4$, which can be written as $\lim_{x \to 2} f(x) = 4$.

Based on the above, we can define that a function $f(x)$ has a limit of 4 as x approaches 2

$$\lim_{x \to 2} f(x) = 4$$

when the following conditions are satisfied:

$$\lim_{x \to 2^-} f(x) = 4 \text{ and } \lim_{x \to 2^-} f(x) = 4$$

The limit of $f(x)$ as x approaches a from the left, $\lim_{x \to a^-} f(x)$, or a from the right, $\lim_{x \to a^+} f(x)$, is referred to as a **one-sided limit**. Consequently, we can conclude that the two-sided limit, $\lim_{x \to a} f(x)$, exists when the results of the one-sided limit are equal, and its mathematical expression is the following:

$$\lim_{x \to a} f(x) = L$$

This expression is valid when the following condition is satisfied.

$$\lim_{x \to a^-} f(x) = L \text{ and } \lim_{x \to a^+} f(x) = L$$

Can we find examples of limits in our daily lives? Are there situations where limits do not exist or where only one-sided limits are present? The answer to both questions is yes, and these situations occur more often than we might realize. Let's explore the concepts of continuity and discontinuity.

3.2 When Functions Are Continuous

Suppose you go to the airport to pick up your friend. The airport parking fee is $10 for the first hour and increases by $5 per hour, as shown in the graph below. The parking fees are $10 from 0 to 59 minutes and 59 seconds. However, after one hour, the cost increases to $15, discontinuing the previous value of $10.

In the graph below, it is continuous only during a certain period, and the continuation of that section uses a line to connect points. There are two types of circles: a closed circle (•, which includes the number) and an open circle (○, which does not include the number). Therefore, in the interval from 0 to 1, the 0 hour is included, but the 1 hour is not.

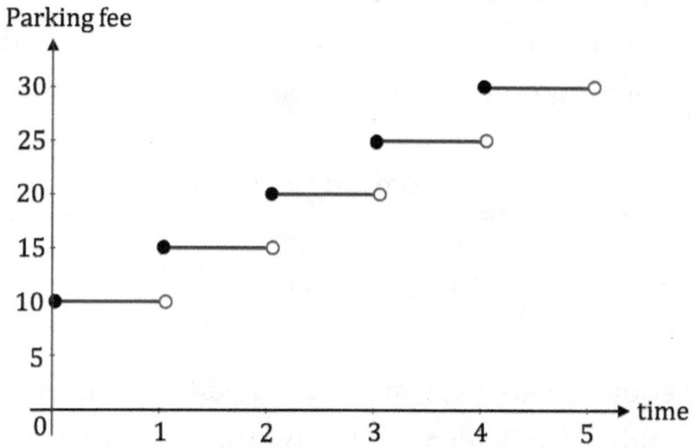

After waiting for your friend for about an hour and paying the parking fee, if the moment you insert your card into the parking machine is 59 minutes and 59 seconds, you will pay a parking fee of $10. However, if you use it at exactly one hour or later, you will be charged $15. There is only a one-second difference between 59 minutes and 59 seconds and one hour, but the rates differ drastically. The reason for this **discontinuity** is that when one second is divided into infinite segments, two connected points are cut off in a very short period of time, causing the parking fee to suddenly change from $10 to $15. This phenomenon is called a **jump discontinuity** in mathematics.

The definition of a **continuous function** is that the output values change continuously as the inputs change, **without any holes or breaks due to sudden jumps from one value to another**. Therefore, the graph above shows discontinuity every hour, such as when t is $1, 2, 3,$ and 4. This occurs because the outputs suddenly jump without taking on the values in between when the inputs are continuously changed.

In the graph above, you indicate that the parking fee has different values when approaching from the left side of 1 and the right side of 1. This situation can be expressed as follows:

$$\lim_{\text{time}\to 1^-} \text{parking fee} = \$10$$

$$\lim_{\text{time}\to 1^+} \text{parking fee} = \$15$$

In the equation above, time is denoted by the input variable t, and the parking fee is represented by the output value $f(t)$. The equation can be rearranged as follows.

$$\lim_{x\to 1^-} f(t) = 10$$

$$\lim_{x\to 1^+} f(t) = 15$$

The equations above tell us that the limit $f(t)$ does not have the same value when x approaches 1 from the left and the right. Consequently, the two-sided limit of $f(t)$ does not exist since $\lim_{t\to 1^-} f(t)$ and $\lim_{t\to 1^+} f(t)$ are not equal due to the discontinuity at $t = 1$. The **discontinuity of a function** at a point $t = 1$ is expressed as follows:

$$\lim_{t\to 1^-} f(t) \neq \lim_{t\to 1^+} f(t)$$

and

$$\lim_{t \to 1} f(t) = DNE$$

Even if the parking fee is not continuous over time, we still have one question. "How much is the parking fee when the clock is exactly one hour?" The question asks whether it is right to pay a parking fee of \$10 or \$15 when the time is exactly one hour.

The parking fee depends on how the airport defines it. The staff may determine that the parking fee will be \$10 if punched exactly one hour into the parking machine. However, the staff sets the fee at \$15 for 1 hour at the parking machine, which will result in a price of \$15. In mathematics, the cost entered by staff for exactly one specific time is called a **defined value** (or **exact value**) and is expressed as follows:

$$f(1) = 15$$

This means that the parking fee for exactly one hour is \$15. In the graph, the value at each hour is represented by a closed circle.

A function might not be continuous even if it satisfies the condition for $\lim_{x \to a^-} f(x) = \lim_{x \to a^+} f(x)$, as shown in the following example. The graph below represents a removable discontinuity, a point on a graph that is not defined. The input $x = 1$ has an output of -1, which is not included in the function itself. Thus, the function $f(x)$ has a hole at $x = 1$, and the function $f(x)$ is discontinuous at $x = 1$, even though the two-sided limit exists as $\lim_{x \to 1^-} f(x) = \lim_{x \to 1^+} f(x)$.

The function above can be expressed as follows:

$$\lim_{x \to 1} f(x) = 1 \text{ since } \lim_{x \to 1^-} f(x) = \lim_{x \to 1^+} f(x) = 1$$

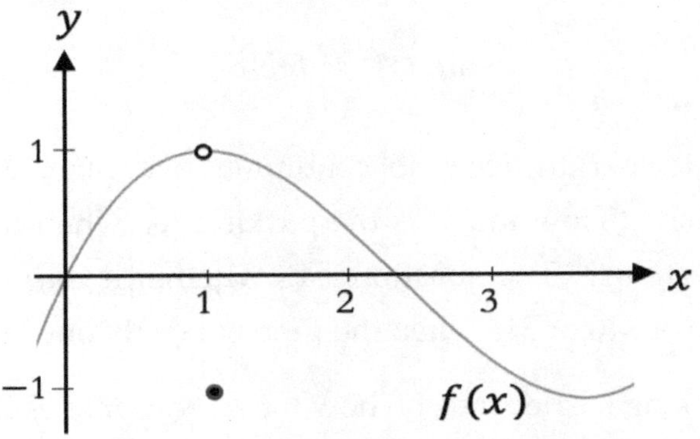

but

$$f(1) = -1$$

Therefore,

$$\lim_{x \to 1} f(x) \neq f(1).$$

The function above demonstrates a **removable discontinuity** — the limit exists at that point, but the function is not defined there. It is expressed as:

$$\lim_{x \to a} f(x) = L \text{ but } f(a) \neq L$$

A popular luxury car brand can illustrate the analogy for a removable discontinuity in our daily lives. This company has never offered discounts on its cars since its launch. However, to celebrate its 100th anniversary, the company plans to sell cars at a 50% discount for just one day. On this special day, the vehicles will be priced at their discounted value. As a result, the price of the vehicles on that day will differ from expectations, creating a temporary disruption in their usual pricing continuity.

The concept of removable discontinuity helps us understand how to find the mathematical expression for a function to be continuous. A function $f(x)$ can be continuous at a point $x = 1$ when the $\lim_{x \to 1} f(x)$ is equal to the function value at that point, as shown in the diagram below. The function $f(x)$ is to be continuous at $x = 1$ since the hole is closed by the same defined value $f(1) = 1$ and the limit value $\lim_{x \to 1} f(x) = 1$. Thus, the condition for function continuity can be expressed as follows:

$$\lim_{x \to 1} f(x) = f(1)$$

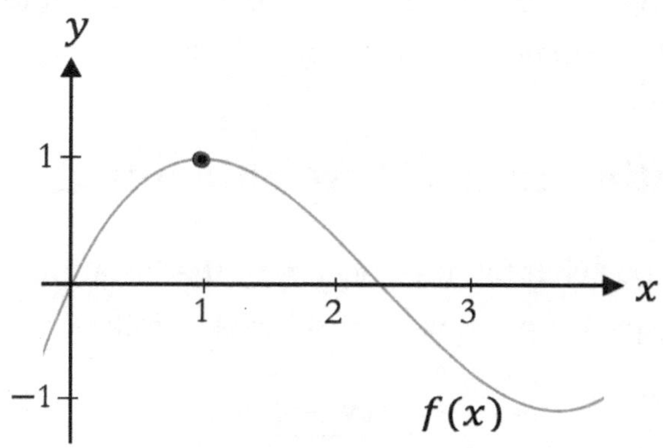

In summary, a function $f(x)$ can be continuous without any breaks or holes. The **continuity of a function at a point** a must satisfy the following conditions:

1. The function is defined at $x = a$:

$$f(a) = L$$

2. The limit exists at $x = a$:

$$\lim_{x \to a^-} f(x) = \lim_{x \to a^+} f(x)$$

3. The limit and defined value must be the same:

$$\lim_{x \to a} f(x) = f(a)$$

Continuity plays a crucial role in calculus. More specifically, continuity allows us to define and compute essential functions that describe real-world phenomena such as motion behaviors, population growth, heat flow, and voltage variation. Because these functions are continuous, we can use limits to precisely define their instantaneous rates of change, which makes it possible to analyze and model how quantities change in calculus.

3.3 Continuity and Differentiability

In the previous chapter, the concept of the instantaneous rate of change at a specific point was introduced as follows:

$$\lim_{x \to a} \frac{f(x) - f(a)}{x - a}$$

The formula above combines the limit with the idea of the average rate of change to find the slope of a tangent line at a specific point $x = a$ for a function $f(x)$. To compute this slope, the function must satisfy certain conditions at $x = a$. First, the function must be defined at $x = a$. Second, the limit of the formula must exist and be finite. These conditions are the same as the continuity of a function introduced in the previous section. When both conditions are satisfied, its value is the slope of the tangent line at $x = a$ for $f(x)$. This is called **differentiability** at $x = a$.

It is important to emphasize the link between continuity and differentiability since it is a key rule in calculus. Differentiability always implies continuity. However, the reverse is not always true: a function can be continuous without being differentiable. For example, a function with a sharp corner or cusp may be continuous (without breaks or holes), but it does not have a well-defined tangent slope at that point, so it is not differentiable there. This distinction is important because when we use limits to define instantaneous rates of change, we need both smoothness and continuity for the tangent slope to exist.

3.4 Checkpoint 3

The formula below shows the payment based on the hours the plumber worked. Adam thinks he must pay \$190 for 1 hour and 30 minutes of work.

$$f(t) = 60t + 100$$

$$f(1.5) = 60(1.5) + 100 = 190$$

The plumber clarified that he charges a flat rate of \$100 for the first hour. After the first hour, he charges an additional \$60 per hour. Thus, \$160 will be charged for 1 to 2 hours of work, and \$220 will be the fee for 2 to 3 hours of work. Adam realized that his linear function did not reflect this stepwise pricing and updated the graph to describe the actual fee structure accurately.

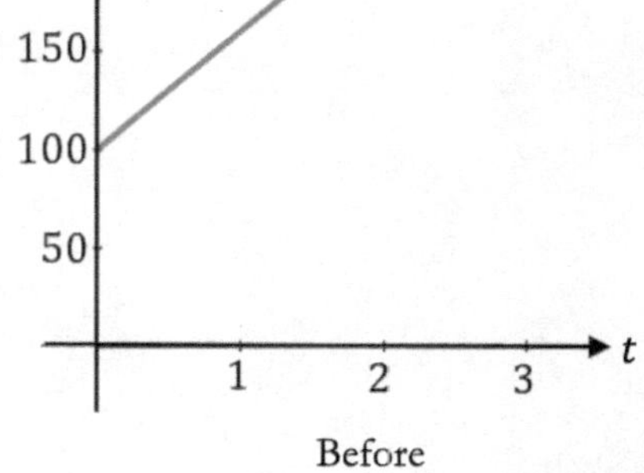

Before

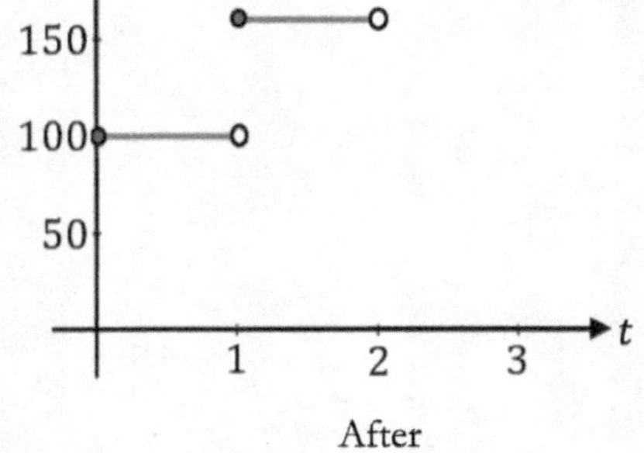

After

1. In the graph of $f(t)$ that was before updated,
   (a) What are the values $\lim_{t\to 1^-} f(t)$ and $\lim_{t\to 1^+} f(t)$?
   (b) What are the values $\lim_{t\to 1} f(t)$ and $f(1)$?
   (c) Why are the values $\lim_{t\to 1} f(t)$ and $f(1)$ the same?
2. In the graph of $g(t)$, which has been updated afterward?
   (a) What are the values $\lim_{t\to 1^-} g(t)$ and $\lim_{t\to 1^+} g(t)$?
   (b) What are the values $\lim_{t\to 1} g(t)$ and $g(1)$?
   (c) Why are the values $\lim_{t\to 1} g(x)$ and $g(1)$ different?
3. Adam believes the plumber would earn a higher wage based on the wage calculated from the left graph. To demonstrate this, he performs the following calculation:
   (a) What is the value of $\lim_{t\to 1.5} f(t)$?
   (b) What is the value of $\lim_{t\to 1.5} g(t)$?
   (c) Which graph represents a higher value?

After calculating, Adam confirmed that the plumber earns more money, as shown in the left graph. He then explained his calculations step by step to the plumber. The graph on the left is more effective for determining a plumber's wage, which is calculated using the formula $g(t) = 60t + 100$. The plumber thanked Adam and suggested that he study mathematics.

3.5 Checkpoint 3 Solution

1. (a) When we approach $t = 1$ from the left side ($t \to 1^-$), the output value of the function gets closer to 160, but $t = 1$ from the right side ($t \to 1^+$), the output value also gets closer to 160. Therefore, $\lim_{t\to 1^-} f(t) = 160$ and $\lim_{t\to 1^+} f(t) = 160$.

   (b) Since the left-hand and right-hand limits are the same, the overall limit as t approaches 1 exists and equals 160. The function is also given to have $f(1) = 160$. So,

   $$\lim_{t\to 1} f(t) = 160 \text{ and } f(1) = 160.$$

   (c) The value of the limit and the actual value of the function at $t = 1$ are the same, which means there is no jump or hole in the graph at that point. This is exactly the definition of continuity at a point:

   $$\lim_{t\to 1} f(t) = f(1)$$

   Therefore, $f(t)$ is continuous at $t = 1$.

2. (a) $\lim_{t\to 1^-} g(t) = 100$ and $\lim_{t\to 1^+} g(t) = 160$.

   (b) The two-sided limit does not exist at $t = 1$ because the left-hand and right-hand limits are not equal. So, $\lim_{t\to 1} g(t) = \text{DNE}$. However, the function value at $t = 1$ is $g(1) = 160$.

   (c) Because the function's defined value at $t = 1$ does not match the limit as t approaches 1. There is a jump discontinuity.

3. (a) Both $\lim_{t\to 1.5^-} f(t)$ and $\lim_{t\to 1.5^+} f(t)$ are equal to 190, so the value of $\lim_{t\to 1.5} f(t) = 190$.

   (b) The graph $g(t)$ is continuous between 1 and 2 since two-sided limit exists at $t = 1.5$ as 160:

$$\lim_{t\to 1.5^-} g(t) = \lim_{t\to 1.5^+} g(t)$$

   (c) The wage calculated using the function $f(t)$ for 1.5 hours is \$190, while the function $g(t)$ gives \$160. Therefore, the plumber should base his wage on the salary model described by $f(t)$.

Part II
Introduction to the Derivative

Chapter 4

Instantaneous Rate of Change

Around us, change is easily found everywhere; it shapes the movement of planets, the rise and fall of stock prices, and even the way our hearts beat. But how do we measure change mathematically? This was our very first question in Chapter 1. In previous discussions, we explored how linear functions change constantly, while nonlinear functions vary unpredictably. To understand this variation, we should review and discuss more the two key ideas for change: the average rate of change and **instantaneous rate of change**.

The average rate of change gives a general overview over an interval. However, to find the precise rate at a specific point, we need a more accurate tool — limits, which connect average and exact measurements — leading us to the concept of instantaneous change. In this chapter, we will expand on this foundation of instantaneous change in both linear and nonlinear functions.

4.1 Understanding the Rate of Change

Imagine you are on a road trip. You drive 200 miles in 4 hours, so your average speed is:

$$\frac{200 \text{ miles}}{4 \text{ hours}} = 50 \text{ mph}$$

This shows how fast you traveled on average over the entire trip. However, your actual speed can be higher or lower at any given moment. If you check your speedometer, you might see 60 mph at one moment and 40 mph at another. The reading on your speedometer at any given moment shows your **instantaneous speed**, or how fast you are moving.

In mathematics, the average rate of change describes how a function's value changes over an interval. This is calculated using the formula for the secant line:

$$\frac{f(b) - f(a)}{b - a}$$

This slope does not require a limit, as it simply measures the change over an interval.

On the other hand, the instantaneous rate is defined using limits:

$$\lim_{x \to a} \frac{f(x) - f(a)}{x - a}$$

It tells us how the function is changing at a specific moment or point and helps us understand how a function behaves at a single point, rather than over an entire interval.

4.2 Instantaneous Rate of Change for Linear Functions

Before exploring the instantaneous rate of change for nonlinear functions, it is helpful to first compare the average and instantaneous rates of change in a linear function. This simpler example makes the concepts much easier to understand. For a linear function, the average rate of change is always constant because the slope remains the same throughout every interval. This naturally raises a question: at a single point on a linear function, does the instantaneous rate of change differ from the average rate of change, or are they the same?

To explore this idea further, let's determine the average rate of change of $f(x) = 2x$ using the values from the linear function shown in the table below. Recall that the slope of a secant line — the average rate of change — is expressed as follows.

Input value x	**Output value** $f(x) = 2x$
1	2
2	4
3	6
a	$2a$
b	$2b$

$$\frac{\Delta y}{\Delta x} = \frac{f(b) - f(a)}{b - a}$$

The average rate of change is calculated using two different input values, $a = 1$ and $b = 2$, along with their corresponding output values, $f(1) = 2$ and $f(2) = 4$, from the table. By substituting this information into the formula, the average rate of change is found to be 2.

$$\frac{f(b) - f(a)}{b - a} = \frac{4 - 2}{2 - 1} = \frac{2}{1} = 2$$

Even if we select different input values, such as 1 and 3, with the corresponding outputs $f(1) = 2$ and $f(3) = 6$, the result will remain the same as 2.

The linear function $f(x) = 2x$ shows that the output always doubles the input value because the rate of change along the line is constant and represented by $\frac{\Delta y}{\Delta x} = 2$, as shown below. The average rate of change for linear functions is constant throughout the entire function.

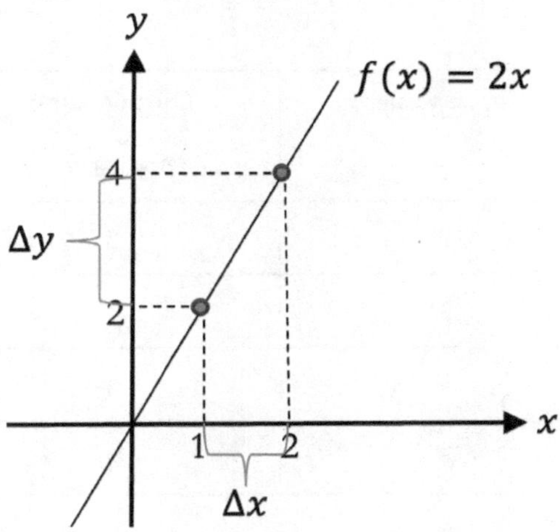

To understand the instantaneous rate of change of this linear function, let's compare it with the average rate of change at $a = 1$ for the function $f(x) = 2x$ using the formula below.

$$\lim_{x \to a} \frac{f(x) - f(a)}{x - a}$$

When calculating the average rate of change for the function $f(x) = 2x$, we choose two fixed input values, such as $a = 1$ and $b = 2$. In contrast, when finding the instantaneous rate of change at $a = 1$, we do not fix a second input. Instead, we examine what happens as x get closer and closer to 1, as shown in the following diagram.

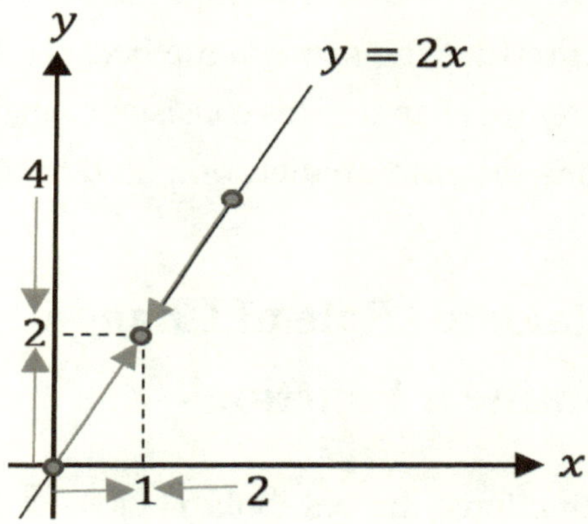

The following procedure introduces how to determine the instantaneous rate of change at $a = 1$ in the function, $f(x) = 2x$.

$$x = x, f(x) = 2x$$

$$a = 1, f(1) = 2$$

Substituting the given values into the formula, we have:

$$\lim_{x \to a} \frac{f(x) - f(a)}{x - a} = \lim_{x \to 1} \frac{2x - 2}{x - 1}$$

This expression becomes $\frac{0}{0}$, which is an indeterminate form. Thus, we should factor and cancel common terms, resulting in 2.

$$\lim_{x \to 1} \frac{2\cancel{(x-1)}}{\cancel{x-1}}$$

$$\lim_{x \to 1} 2 = 2$$

For the linear function $f(x) = 2x$, the average rate of change over any interval and the instantaneous rate of change at a specific point, such as $x = 1$, yield the same result of 2. Although these two rates of change are defined using different mathematical approaches, they produce the same value because linear functions have a constant rate of change. This consistency makes the behavior of linear functions easier to predict compared to nonlinear ones.

4.3 Instantaneous Rate of Change for Nonlinear Functions

Unlike linear functions, the parabola $f(x) = x^2$ does not have a constant slope, so its average and instantaneous rates of change differ. Using the values in the table below, we can examine how these rates of change behave.

The average rate of change at two points $a = 1$ and $b = 2$ of the function at $f(x) = x^2$ can be determined by substituting the given values into the formula below. Then, the following result is obtained:

$$a = 1 \text{ and } f(1) = 1,\ b = 2 \text{ and } f(2) = 4$$

$$\frac{f(b) - f(a)}{b - a} = \frac{4 - 1}{2 - 1} = \frac{3}{1} = 3$$

Input value x	Output value $f(x) = x^2$
1	1
2	4
3	9
4	16
a	a^2
b	b^2

When we substitute other inputs from the table, such as $a = 1$ and $b = 3$, or $b = 4$, into the formula, the results will be 4 and 5 respectively. The calculated results are plotted on the graph of $f(x) = x^2$ below. As b increases to 2, 3, and 4, the average rate of change from the point $a = 1$ also increases to 3, 4, and 5, respectively.

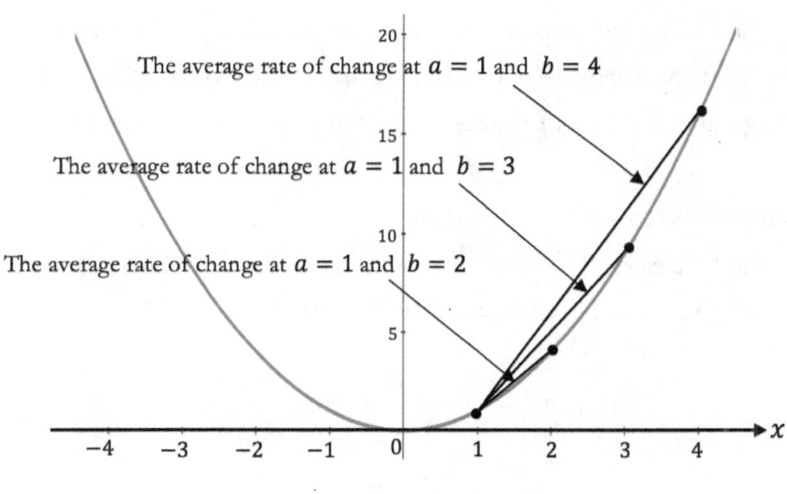

This illustrates that the average rate of change depends on the chosen interval, since it describes the overall change in the function's value over that range. However, the average rate of change does not specify how the function behaves at a single point. Imagine a car moving along the curve $f(x) = x^2$, and we want to find the exact speed at a specific moment rather than the overall trend over an interval. The average rate of change does not provide the precise information we need.

Using the formula below, we can calculate the slope at the single point $a = 1$ for the function $f(x) = x^2$.

$$\lim_{x \to a} \frac{f(x) - f(a)}{x - a}$$

For the function $f(x) = x^2$, as x approaches 1, the corresponding function value $f(x) = x^2$ approaches $f(1) = 1$. However, it is crucial to recognize that x remains a variable — it cannot be equal to 1 but instead continually gets closer to $a = 1$. This distinction allows us to compute the slope of the tangent line at that single point.

The following step-by-step mathematical derivation illustrates how the limit formula is used to compute the instantaneous rate of change and how it compares to the average rate of change.

1. Apply the formula to the function.
For the function $f(x) = x^2$, we want to find the instantaneous rate of change at $a = 1$. After substituting into the formula, we have:

$$\lim_{x \to a} \frac{f(x) - f(a)}{x - a} = \lim_{x \to 1} \frac{x^2 - 1}{x - 1}$$

2. Simplify the expression and cancel the common terms.
As you notice, the numerator $x^2 - 1$ is a difference of squares and can be factored:

$$(x^2 - 1) = (x - 1)(x + 1)$$

Substituting this back into the limit, we have:

$$\lim_{x \to 1} \frac{(x + 1)(x - 1)}{x - 1}$$

Since x approaches 1, and the limit is an indeterminate form $\frac{0}{0}$, we can cancel the common factor $(x - 1)$ from the numerator and the denominator.

$$\lim_{x \to 1} \frac{(x + 1)\cancel{(x - 1)}}{\cancel{x - 1}}$$

3. Evaluate the limit. After simplifying the expression, we can directly substitute $x = 1$ into $(x + 1)$ because it is no longer an indeterminate form and the function is continuous at that point.

$$\lim_{x \to 1} (x + 1) = 2$$

The calculation above can be summarized as follows:

$$\begin{aligned} \lim_{x \to a} \frac{f(x) - f(a)}{x - a} &= \lim_{x \to 1} \frac{x^2 - 1}{x - 1} \\ &= \lim_{x \to 1} \frac{(x + 1)\cancel{(x - 1)}}{\cancel{x - 1}} \\ &= \lim_{x \to 1} (x + 1) \\ &= 2 \end{aligned}$$

Using the same process, we can determine the instantaneous rate of change at $a = 2$ and $a = 3$ as follows:

$$a = 2 \text{ and } f(2) = 4, x = x \text{ and } f(x) = x^2$$

$$\begin{aligned}
\lim_{x \to a} \frac{f(x) - f(a)}{x - a} &= \lim_{x \to 2} \frac{x^2 - 4}{x - 2} \\
&= \lim_{x \to 2} \frac{(x+2)\cancel{(x-2)}}{\cancel{x-2}} \\
&= \lim_{x \to 2} (x + 2) \\
&= 4
\end{aligned}$$

$$a = 3 \text{ and } f(3) = 9, x = x \text{ and } f(x) = x^2$$

$$\begin{aligned}
\lim_{x \to a} \frac{f(x) - f(a)}{x - a} &= \lim_{x \to 3} \frac{x^2 - 9}{x - 3} \\
&= \lim_{x \to 3} \frac{(x+3)\cancel{(x-3)}}{\cancel{x-3}} \\
&= \lim_{x \to 3} (x + 3) \\
&= 6
\end{aligned}$$

The instantaneous rates of change calculated earlier are plotted on the function $f(x) = x^2$ below. The slopes of the tangents increase to 2, 4, and 6, respectively, as the values of x change to 1, 2, and 3.

This distinction is crucial: while **the average rate of change is calculated over an interval, the instantaneous rate of change is determined by examining the function's behavior at a specific point using limits.**

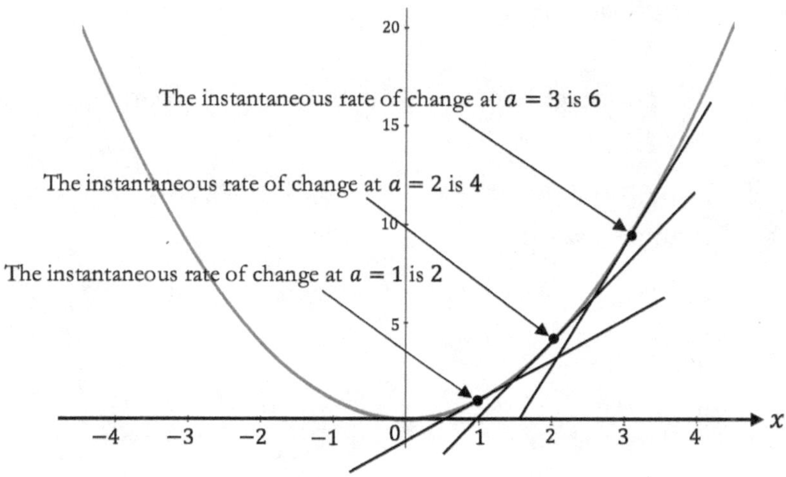

In other words, we can say that the instantaneous rate of change provides an accurate description of how the function varies at any point along the curve.

The benefits of this method apply across many fields in science, engineering, and business. For instance, determining the instantaneous rate of change can precisely forecast infection rates during a pandemic. It can also assess how the growth rate of cancer cells responds to different treatments. When designing a bridge, changes in weight distribution caused by variable traffic from cars or pedestrians can be calculated or simulated. Additionally, it can predict how shifts in product prices will influence consumer purchasing habits. The applications of this mathematical approach are virtually limitless.

4.4 Checkpoint 4

Adam kicked a soccer ball, and its path followed a parabolic shape. The height of the ball over time is modeled by the function $h(t) = 4t - t^2$. This function represents how the height of the ball (in meters) changes over time t (in seconds).

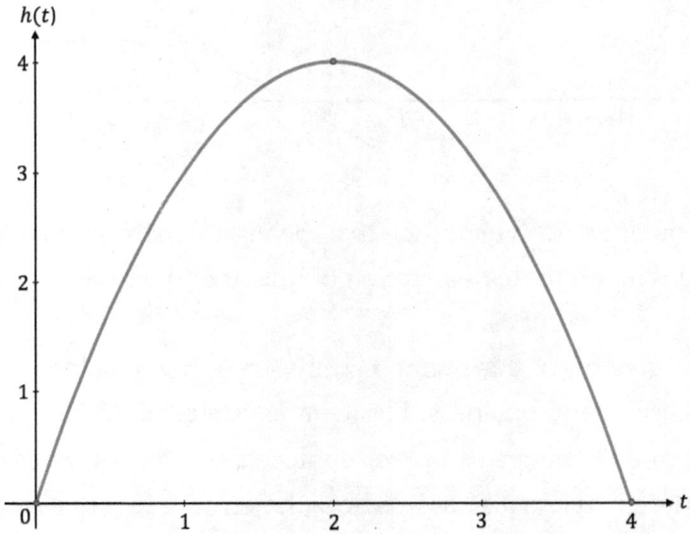

Using the formula below:

- **Average rate of change**

$$\frac{f(x) - f(a)}{x - a}$$

- **Instantaneous rate of change**

$$\lim_{x \to a} \frac{f(x) - f(a)}{x - a}$$

Answer the following questions.

1. Calculate the average rate of change of the function $h(t) = 4t - t^2$ over the time interval from 0 to 4 seconds.

   (a) What does this result tell us about the motion of the ball?

   (b) How should we interpret this in the context of the rate of change?

2. Find the instantaneous rate of change for the function $h(t) = 4t - t^2$ at $t = 2$.

   (a) What does this value represent in the context of the ball's movement?

   (b) The average rate of change in question (a) reflects the overall change from launch to landing, such as from ground to ground. In contrast, why is the instantaneous rate of change at $t = 2$ equal to 0? What does this tell us about the ball's height at that exact moment?

3. Although both questions above may produce the same numerical result, they describe different situations. Explain how the average and instantaneous rate of change provide various insights into the ball's motion.

4.5 Checkpoint 4 Solution

1. We use the formula:

$$\frac{f(x) - f(a)}{x - a}$$

So,

$$\begin{aligned}
\frac{h(4) - h(0)}{4 - 0} &= \frac{(4 \cdot 4 - 4^2) - (4 \cdot 0 - 0^2)}{4} \\
&= \frac{(16 - 16) - 0}{4} \\
&= \frac{0}{4} \\
&= 0
\end{aligned}$$

   (a) The result shows that the ball's height did not change between $t = 0$ and $t = 4$. The average rate of change is 0, since the ball started and ended at ground level (height = 0).

   (b) Because the ball started and ended at the same height (the ground), the net change in height is 0, even though it went up and came back down.

2. We use the formula

$$\lim_{x \to a} \frac{f(x) - f(a)}{x - a}$$

$$\begin{aligned}
\lim_{t\to 2}\frac{h(t)-h(2)}{t-2} &= \lim_{t\to 2}\frac{(4t-t^2)-(4\cdot 2-2^2)}{t-2} \\
&= \lim_{t\to 2}\frac{4t-t^2-4}{t-2} \\
&= \lim_{t\to 2}\frac{-(t^2-4t+4)}{t-2} \\
&= \lim_{t\to 2}\frac{-(t-2)\cancel{(t-2)}}{\cancel{t-2}} \\
&= \lim_{t\to 2}-(t-2) \\
&= 0
\end{aligned}$$

So, at $t = 2$, the instantaneous rate of change is 0.

(a) The instantaneous rate of change is 0, which means that at exactly 2 seconds, the ball reached its peak height and took a temporary pause in its upward motion before starting to fall downward.

(b) The slope of the tangent line is 0 at $t = 2$, indicating the highest point of the ball's path. The instantaneous rate of change (the ball's velocity) is 0 at that moment, marking the turning point from going up to coming down.

3. The average rate of change from $t = 0$ to $t = 4$ and the instantaneous rate at $t = 2$ are both 0, but they mean very different things. The 0 average rate of change indicates that the ball's overall height didn't change during the entire flight (from start to finish). However, the instantaneous rate of change being 0 means that the ball paused at the peak point momentarily.

Chapter 5

Derivative

In our exploration of change, we saw how limits allow us to determine a function's instantaneous rate of change, such as how an object moves at a specific moment. But how can we extend this idea to any function at any point? This question introduces us to one of the most powerful concepts in calculus: the **derivative**.

The derivative provides **a systematic way to find the slope of a tangent line** and **measure how a function changes at an exact point**. In this chapter, we will examine the concept of the derivative, its formal definition, and its importance in analyzing dynamic systems. Mastering this concept opens the door to understanding and predicting change in physics, engineering, economics, and many other fields.

5.1 Derivatives Explained Step by Step

The formula for the instantaneous rate of change describes how to calculate the instantaneous rate of change for the function $f(x) = x^2$ at the point $a = 1, 2$, or 3.

$$\lim_{x \to a} \frac{f(x) - f(a)}{x - a}$$

We can find the instantaneous rate of change of the function $f(x) = x^2$ at $a = 1$ and $f(1) = 1$ after plunging into the formula as follows:

$$\lim_{x \to 1} \frac{x^2 - 1}{x - 1} = \lim_{x \to 1} \frac{(x+1)\cancel{(x-1)}}{\cancel{x-1}} = \lim_{x \to 1} (x+1) = 2$$

and at $a = 2$ and $f(2) = 4$

$$\lim_{x \to 2} \frac{x^2 - 4}{x - 2} = \lim_{x \to 2} \frac{(x+2)\cancel{(x-2)}}{\cancel{x-2}} = \lim_{x \to 2} (x+2) = 4$$

and at $a = 3$ and $f(3) = 9$

$$\lim_{x \to 3} \frac{x^2 - 9}{x - 3} = \lim_{x \to 3} \frac{(x+3)\cancel{(x-3)}}{\cancel{x-3}} = \lim_{x \to 3} (x+3) = 6$$

Similarly, the instantaneous rate of change of the function $f(x) = x^2$ with a negative value such as $a = -1, -2$, or -3 can be calculated. At $a = -1$ and $f(-1) = 1$ as follows:

$$\lim_{x \to -1} \frac{x^2 - 1}{x + 1} = \lim_{x \to -1} \frac{\cancel{(x+1)}(x-1)}{\cancel{x+1}} = \lim_{x \to -1} (x-1) = -2$$

and at $a = -2$ and $f(-2) = 4$

$$\lim_{x \to -2} \frac{x^2 - 4}{x + 2} = \lim_{x \to -2} \frac{\cancel{(x+2)}(x-2)}{\cancel{x+2}} = \lim_{x \to -2} (x-2) = -4$$

and at $a = -3$ and $f(-3) = 9$

$$\lim_{x \to -3} \frac{x^2 - 9}{x + 3} = \lim_{x \to -3} \frac{\cancel{(x+3)}(x-3)}{\cancel{x+3}} = \lim_{x \to -3} (x-3) = -6$$

The calculation above reveals a recurring pattern. In mathematics, such patterns can be generalized to simplify computations. Based on this pattern, we can quickly develop a method to determine the instantaneous rate of change.

In the process described, the formula replaced the variable a with various positive and negative numbers. To avoid repetitive calculations, it is more efficient to keep a as a variable rather than substituting specific values each time.

$$\begin{aligned}\lim_{x\to a}\frac{f(x)-f(a)}{x-a} &= \lim_{x\to a}\frac{x^2-a^2}{x-a}\\ &= \lim_{x\to a}\frac{(x+a)\cancel{(x-a)}}{\cancel{x-a}}\\ &= \lim_{x\to a}(x+a)\end{aligned}$$

What does this result mean when it includes a instead of a specific number? In the above expression, $\lim_{x\to a}(x+a)$ can be separated into two parts. One is the $\lim_{x\to a} x$ expressed as the limit for $f(x) = x$, and the other is $\lim_{x\to a} a$ the limit for $f(x) = a$. The result for $\lim_{x\to a} x$ is a since the function $f(x) = x$ increases proportionally as x increases, as shown in the graph below on the left.

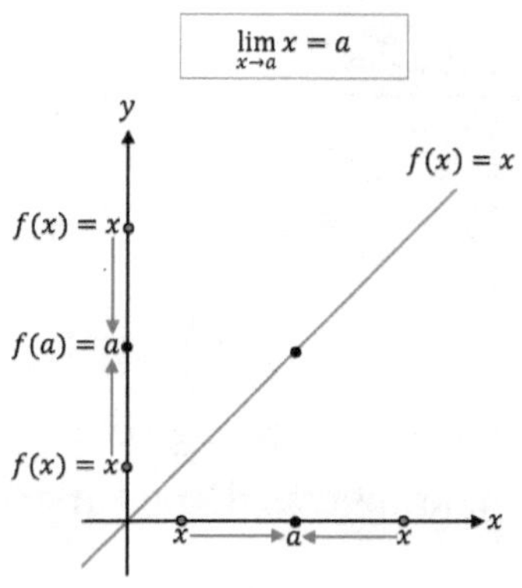

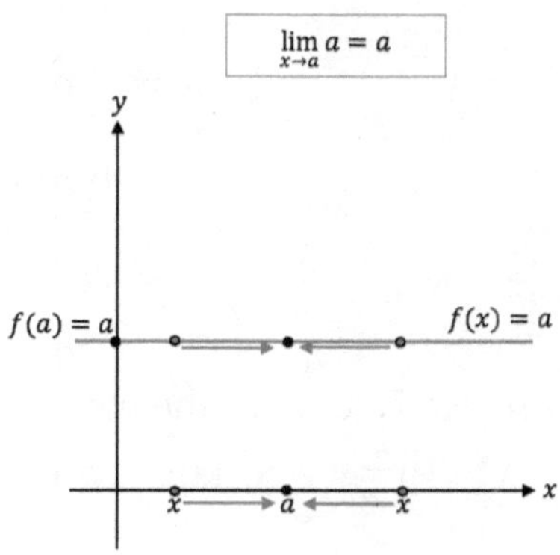

However, the result for $\lim_{x\to a}$ is a, because the results of the limit of a constant function $f(x) = a$ always give the constant value a as output, regardless of the value of x, as shown in the graph on the right. Hence, the result of $\lim_{x\to a}(x+a)$ becomes $2a$.

After applying the distribution of the limit sign based on the result above, we can find the same result as follows:

$$\begin{aligned}\lim_{x\to a}(x+a) &= \lim_{x\to a} x + \lim_{x\to a} a \\ &= a + a \\ &= 2a\end{aligned}$$

The significance of this value, $2a$, is that it allows us to find the instantaneous rate of change of $f(x) = x^2$ at any input $x = a$ without repeating the calculation each time.

We use the notation f' (pronounced "f **prime**") to represent the derivative of the function f with respect to a as following:

$$\begin{aligned}f'(a) &= \lim_{x\to a} \frac{f(x) - f(a)}{x - a} \\ &= \lim_{x\to a} \frac{x^2 - a}{x - a} \\ &= \lim_{x\to a} \frac{(x+a)\cancel{(x-a)}}{\cancel{x-a}} \\ &= \lim_{x\to a}(x+a) \\ &= 2a\end{aligned}$$

This process calculates the instantaneous rate of change of the function $f(x) = x^2$ at a certain input $x = a$, known as **differentiation**.

The $f'(a)$ represents the derivative of $f(x) = x^2$ at a specific point a on the function. Using the result $f'(a) = 2a$, we can easily find the instantaneous rate of change of $f(x) = x^2$ at any point, such as $a = 1, 2, 3, -1, -2,$ and -3 without performing complicated calculations repeatedly.

$$f'(a) = 2a$$

For example, if $a = 0$, then $f'(0) = 2(0) = 0$, and if $a = 1$, then $f'(1) = 2(1) = 2$, and so on as shown below.

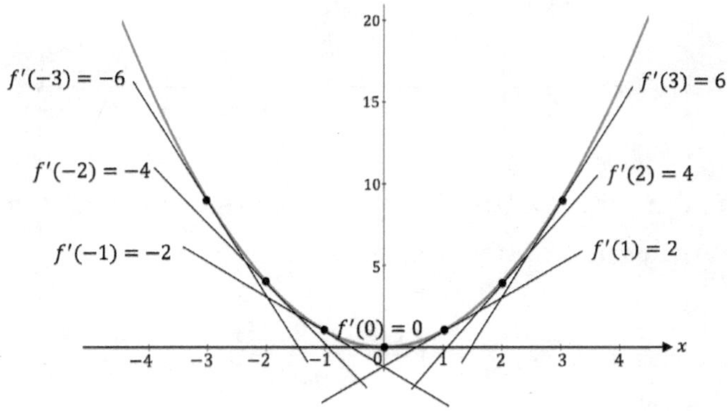

By using the derivative of a function, we can easily identify the instantaneous rate of change at any given input value. For this reason, the derivative is a mathematical tool for calculating the instantaneous rate of change.

5.2 Standard Definition of the Derivative

We can derive an alternative derivative formula if the x is h units away from a, as shown in the figure below. As the distance h between a and x decreases and approaches zero, the secant line between $(a, f(a))$ and $(a+h, f(a+h))$ gets closer to and approaches the tangent line. This form highlights the idea of how the function behaves when we move a small distance, h, along the input from a. It is more practical for calculations and more convenient for proofs.

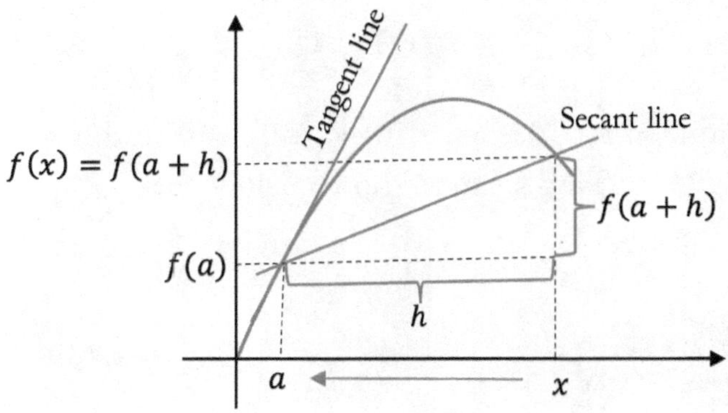

In the derivative equation below, we can replace x with $a+h$ and then the equation transforms like this.

$$\begin{aligned} f'(a) &= \lim_{x \to a} \frac{f(x) - f(a)}{x - a} \\ &= \lim_{a+h \to a} \frac{f(a+h) - f(a)}{(a+h) - a} \end{aligned}$$

The sign $x \to a$ in limit notation can be replaced with $a+h \to a$ since we can substitute $a+h$ for x. By subtracting a from both sides of the expression, we obtain $a+h-a \to a-a$, which simplifies to $h \to 0$. Consequently, we arrive at the following derivative formula:

$$f'(a) = \lim_{h \to 0} \frac{f(a+h) - f(a)}{h}$$

The formula above is not just an alternative method for finding the derivative; it is **the standard definition of the derivative**. This form is widely used because it allows us to represent the derivative as a general function of x, making it more practical for calculations and for establishing the rules of differentiation. Using this formula, let's calculate the derivative of $f(x) = x^2$. Applying the formula above with $f(a+h) = (a+h)^2$ and $f(a) = a^2$, we have:

$$\begin{aligned} f'(a) &= \lim_{h \to 0} \frac{(a+h)^2 - a^2}{h} \\ &= \lim_{h \to 0} \frac{a^2 + 2ah + h^2 - a^2}{h} \\ &= \lim_{h \to 0} \frac{2ah + h^2}{h} \\ &= \lim_{h \to 0} \frac{\not{h}(2a + h)}{\not{h}} \\ &= \lim_{h \to 0} (2a + h) \\ &= 2a \end{aligned}$$

The limit $\lim_{h \to 0}(2a + h)$ is equal to $2a$ since h goes to 0. Therefore, the derivative formula above yields the same result.

5.3 Derivative as a Function

We have derived the **formulas for the derivative** as follows:

$$f'(a) = \lim_{x \to a} \frac{f(x) - f(a)}{x - a}$$

or

$$f'(a) = \lim_{h \to 0} \frac{f(a+h) - f(a)}{h}$$

These formulas serve as mathematical tools for calculating the instantaneous rate of change of $f(x)$ at any given input. By substituting a specific input value into the formula, we obtain the corresponding instantaneous rate of change as output.

The machine analogy called "**Differentiation**," which calculates the instantaneous rate of change for given inputs using the formula above, provides a better understanding. For example, entering 1 produces the output $f'(1)$, while entering a results in $f'(a)$.

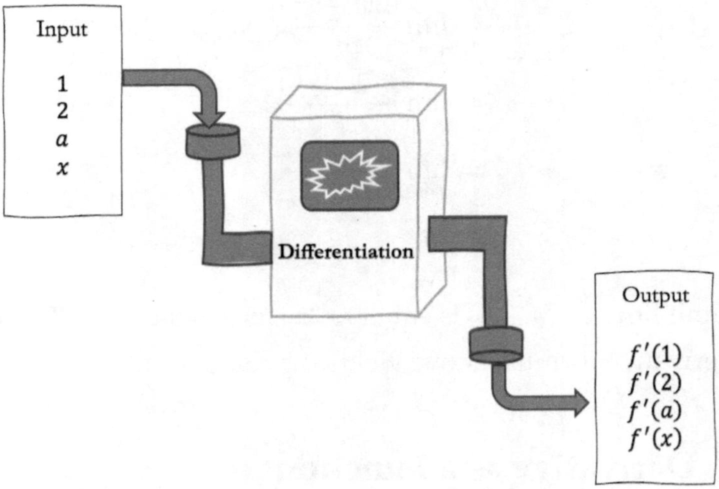

If you substitute a variable x for a in the machine, the output will be $f'(x)$ instead of $f'(a)$. The $f'(x)$ is called the **derivative function** of $f(x)$, and **it demonstrates how the instantaneous rate of change behaves across the entire function** $f(x)$**, whereas** $f'(a)$ **provides the derivative value at a specific point** a. This broader perspective is helpful because it provides a flexible, general way to examine how a function varies throughout, not just at a single point.

The derivative function $f'(x)$ for x is represented as follows:

$$f'(x) = \lim_{h \to 0} \frac{f(x+h) - f(x)}{h}$$

Before discussing the importance of the derivative function, $f'(x)$, let's find the derivative of $f(x) = x^2$ using the derivative function below:

$$\begin{aligned}
f'(x) &= \lim_{h \to 0} \frac{(x+h)^2 - x^2}{h} \\
&= \lim_{h \to 0} \frac{x^2 + 2xh + h^2 - x^2}{h} \\
&= \lim_{h \to 0} \frac{2xh + h^2}{h} \\
&= \lim_{h \to 0} \frac{\cancel{h}(2x+h)}{\cancel{h}} \\
&= \lim_{h \to 0} (2x+h) \\
&= 2x
\end{aligned}$$

The above calculation shows that the derivative function of $f(x) = x^2$ is $f'(x) = 2x$. Using this derivative function, we can easily determine the values of the instantaneous rate of change at $x = -2, x = 0$, and $x = 2$ by plugging these values into the equation $f'(x)$ below.

$$f'(x) = 2x \text{ at any } x$$

Thus, we can calculate the derivative value at any point x using the equation as follows: Instantaneous rate of change at $x = -2$

$$f'(-2) = 2(-2) = -4$$

Instantaneous rate of change at $x = 0$

$$f'(0) = 2(0) = 0$$

Instantaneous rate of change at $x = 2$

$$f'(2) = 2(2) = 4$$

The graphs below illustrate the relationship between $f(x)$ and $f'(x)$. The top graph shows the function $f(x) = x^2$, while the one below presents its derivative, $f'(x) = 2x$.

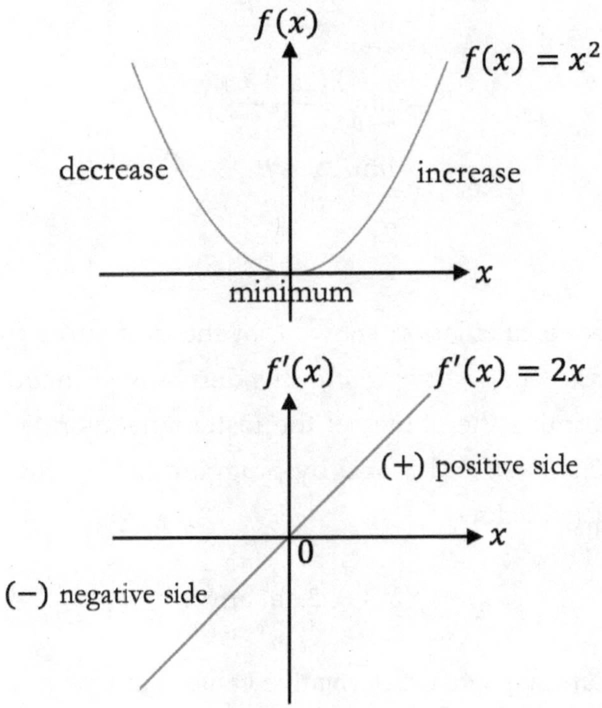

The derivative function $f'(x)$ represents how rapidly the function $f(x)$ changes at any given point. When we compare the two functions vertically, we can clearly identify the relationship and pattern between $f(x)$ and $f'(x)$. This also connects to our earlier observation in Chapter 1 about $\Delta(\Delta y)$, since $f'(x)$ is essentially the graph that represents the slope at every point on $f(x)$.

The graphs show where the function $f(x) = x^2$ **decreases,** which occurs when the derivative function $f'(x) = 2x$ yields a **negative number**. Conversely, the function $f(x) = x^2$ **increases** when the derivative function $f'(x) = 2x$ yields a **positive number**. Additionally, the **minimum value** of the parabola occurs at $f'(x) = 0$.

By using this relationship, we can determine whether the function $f(x)$ will increase or decrease by examining the sign of $f'(x)$, as summarized in the table below. The points at which a function $f(x)$ has **a minimum or maximum** can be determined by finding the value of x for which **the derivative equals zero,** $f'(x) = 0$. This is called a **stationary point** of the function $f(x)$.

$f(x)$	$f'(x)$
Increasing section	(+) positive number
Decreasing section	(−) negative number
Minimum or Maximum	Zero

The following is a real-life example of the relationship between $f(x)$ and $f'(x)$. Adam and Hannah are playing catch with a ball. When Adam throws the ball to Hannah, its height increases as it rises, momentarily stops at its peak, and then falls due to gravity. Typically, the ball's height over time is represented by the position function $s(t)$, where t represents time. The path of the ball follows a parabolic curve, as shown below. What does this tell us about how

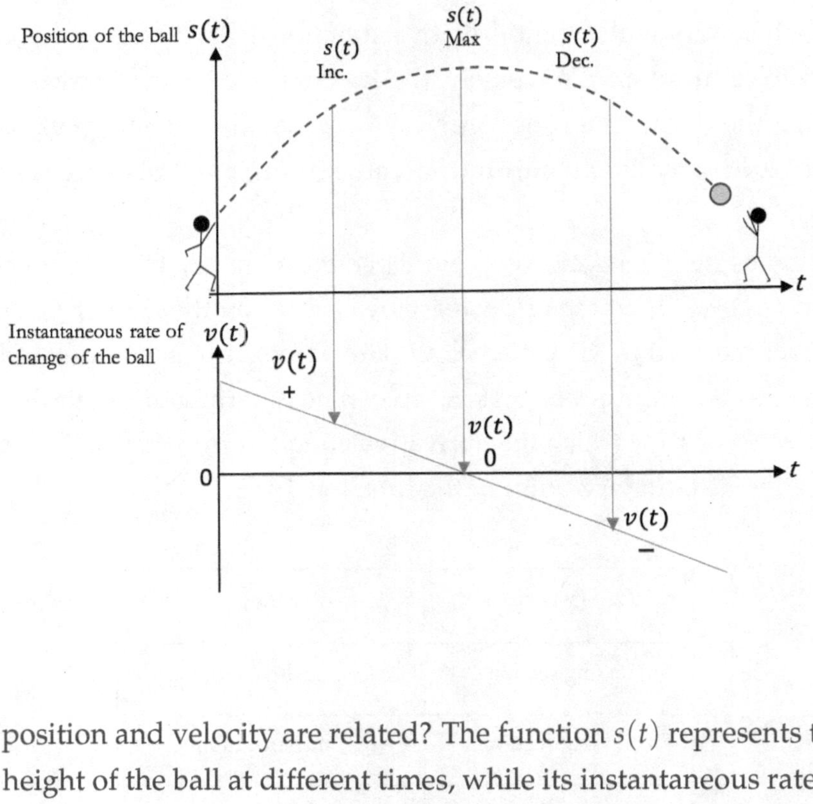

position and velocity are related? The function $s(t)$ represents the height of the ball at different times, while its instantaneous rate of change, $s'(t)$, gives the ball's velocity, $v(t)$, at each moment. When the ball moves upward, the velocity is positive. At the highest point, the ball stops rising for an instant before beginning to fall, so its velocity is zero at that moment. As the ball falls, the velocity becomes negative. The diagrams above illustrate this critical connection between the position function of the ball $s(t)$ and its derivative, the velocity function $v(t)$.

5.4 Common Symbols for Derivatives

The derivative has various mathematical notations depending on the situation. Among them, the most representative are $\frac{dy}{dx}$, similar to the $\frac{\Delta y}{\Delta x}$ of the average rate of change.

As you know, $f(x)$ and y represent a function. Therefore, $f'(x)$ and y' are symbols to describe the derivative of a function. Thus, $\frac{dy}{dx}$ can also be written $\frac{d[f(x)]}{dx}$. The notation shown in the following symbols represents the **first derivative functions**.

$$f'(x) = y' = \frac{dy}{dx} = \frac{d[f(x)]}{dx}$$

5.5 Checkpoint 5

1. Let's determine the derivative function of $f(x)$ using the formula below.

$$f'(x) = \lim_{h \to 0} \frac{f(x+h) - f(x)}{h}$$

   (a) Find the derivative of the function of $f(x) = x^2 + 1$.
   Hint: $f(x+h) = (x+h)^2 + 1$

   (b) Find the derivative of the function of $f(x) = x^2 - 1$.
   Hint: $f(x+h) = (x+h)^2 - 1$

   (c) Based on the results for questions (a) and (b), explain the outcome in your own words.

2. Sketch a graph of the derivative function $f'(x)$ for the provided function $f(x)$.

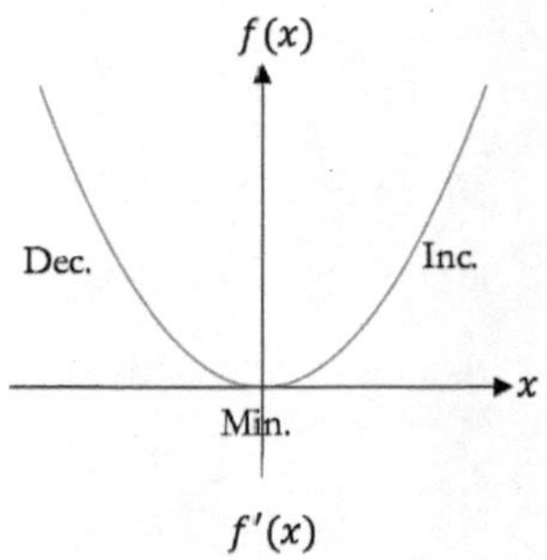

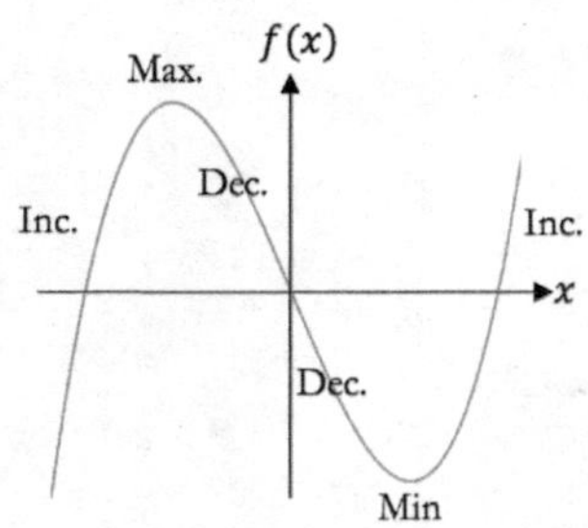

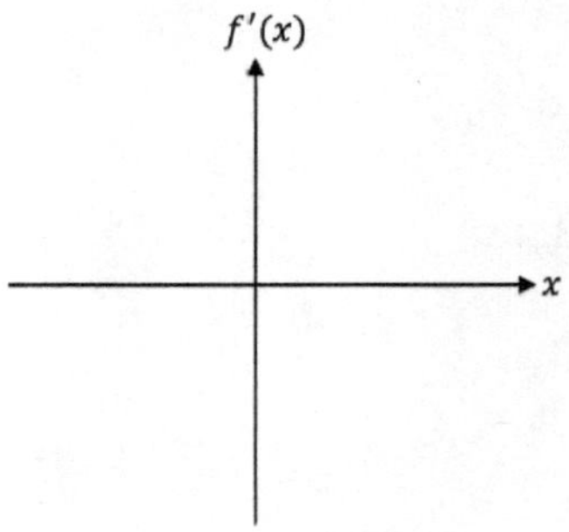

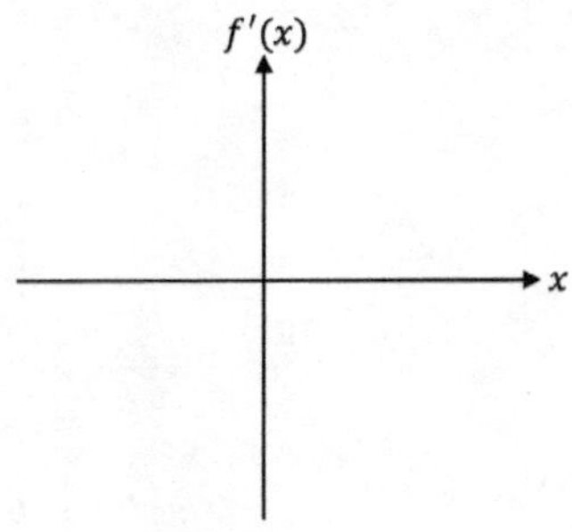

3. The derivative of the position function $s(t)$ is the velocity $v(t)$ such that $s'(t) = v(t)$. Using this relationship, solve the following questions.

   Adam hit a golf ball, and the function represents its trajectory: $s(t) = 9.8t - 4.9t^2$. This function shows the height of the golf ball, meters, as a function of time (t) after being struck, as shown below.

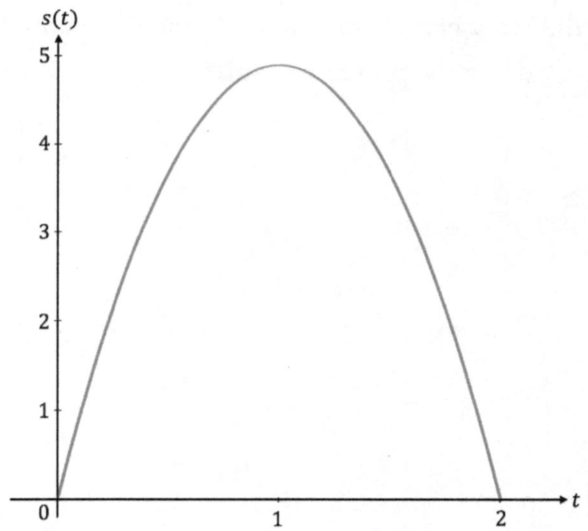

   (a) He wants to determine the average velocity $\bar{v}(t)$ over the interval between $t = 0$ and $t = 1$. Show your steps to calculate the average velocity.

   (b) Use the limit definition of the derivative to find the instantaneous velocity at time $t = a$ from the position function $s(t)$. Show your steps clearly.

   (c) He wants to determine the initial velocity of the ball and the maximum height using the velocity function $v(t)$.

i. What is the initial velocity of the ball at $t = 0$? Explain your method for determining the initial velocity.

ii. The instantaneous velocity of the ball at its maximum height is zero since the ball has a velocity of 0 at the instant that it changes direction from upward to downward. Using the previous result, what is the ball's maximum height?

iii. How can we calculate the ball's maximum height using the previous result?

5.6 Checkpoint 5 Solution

1. (a)

$$\begin{aligned}
f'(x) &= \lim_{h\to 0} \frac{f(x+h)-f(x)}{h} \\
&= \lim_{h\to 0} \frac{[(x+h)^2+1]-(x^2+1)}{h} \\
&= \lim_{h\to 0} \frac{x^2+2xh+h^2+1-x^2-1}{h} \\
&= \lim_{h\to 0} \frac{2xh+h^2}{h} \\
&= \lim_{h\to 0} \frac{\not{h}(2x+h)}{\not{h}} \\
&= \lim_{h\to 0}(2x+h) \\
&= 2x
\end{aligned}$$

(b)

$$\begin{aligned}
f'(x) &= \lim_{h\to 0} \frac{[(x+h)^2-1]-(x^2-1)}{h} \\
&= \lim_{h\to 0} \frac{x^2+2xh+h^2-1-x^2+1}{h} \\
&= \lim_{h\to 0} \frac{2xh+h^2}{h} \\
&= \lim_{h\to 0} \frac{\not{h}(2x+h)}{\not{h}} \\
&= \lim_{h\to 0}(2x+h) \\
&= 2x
\end{aligned}$$

(c) The derivative function $f'(x)$ represents the instantaneous rate of change of $f(x) = x^2$. This rate of change remains the same even if the graph of $f(x)$ is shifted up or down. For example, $f(x) = x^2 + 1$ and $f(x) = x^2 - 1$ both have the same derivative, $f'(x) = 2x$, as shown below. Thus, when we find the derivative, The constant terms do not affect the rate of change because they do not vary with the changes in x. For example, whether we add 1, subtract 1, or add 100, the slope of the curve remains the same.

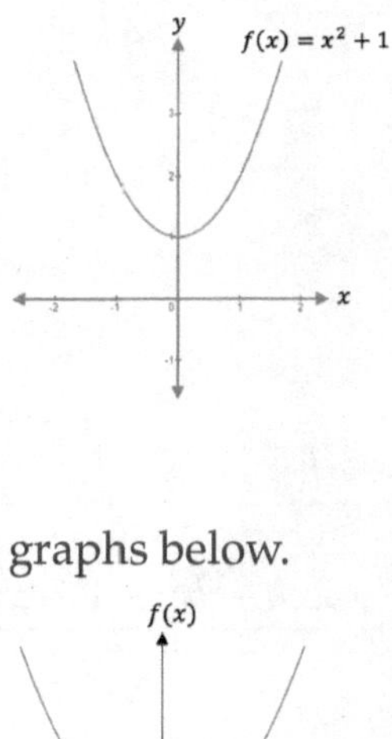

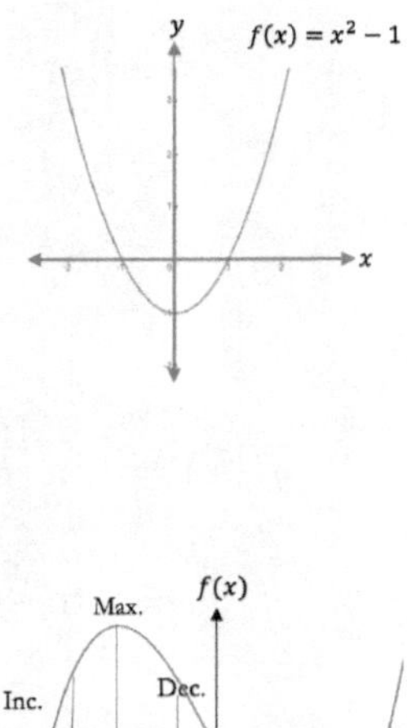

2. See the graphs below.

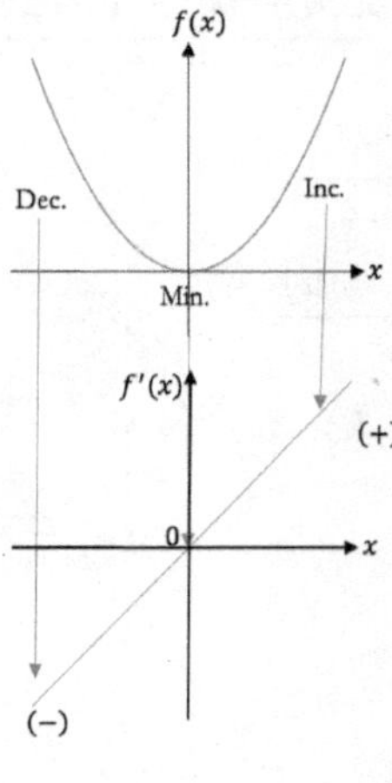

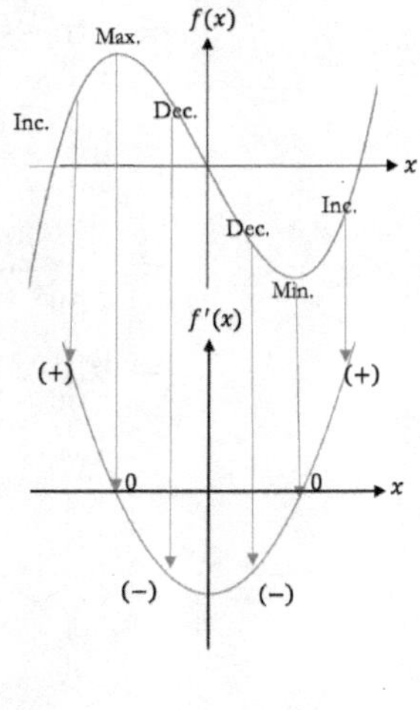

3. (a) To find the average velocity $\bar{v}(t)$ at $t = 0$ and $t = 1$ for $s(t) = 9.8t - 4.9t^2$, we need to substitute the values into the formula below:

$$s(0) = 9.8(0) - 4.9(0)^2 = 0$$

$$s(1) = 9.8(1) - 4.9(1)^2 = 4.9$$

$$\bar{v}(t) = \frac{\Delta s}{\Delta t} = \frac{s(1) - s(0)}{1 - 0} = \frac{4.9 - 0}{1 - 0} = 4.9\ m/sec$$

(b) To find the instantaneous velocity $v(t)$ at $t = a$ from $s(t) = 9.8t - 4.9t^2$, we need to substitute the values into the formula below:

$$v(t) = \lim_{t \to a} \frac{s(t) - s(a)}{t - a}$$

$$\begin{aligned} v(t) &= \lim_{t \to a} \frac{(9.8t - 4.9t^2) - (9.8a - 4.9a^2)}{t - a} \\ &= \lim_{t \to a} \frac{9.8(t - a) - 4.9(t^2 - a^2)}{t - a} \\ &= \lim_{t \to a} \frac{9.8(t - a) - 4.9(t - a)(t + a)}{t - a} \\ &= \lim_{t \to a} \frac{9.8\cancel{(t - a)} - 4.9\cancel{(t - a)}(t + a)}{\cancel{t - a}} \\ &= \lim_{t \to a} 9.8 - 4.9(t + a) \\ &= 9.8 - 4.9(2a) \\ &= 9.8 - 9.8a \\ &= 9.8(1 - a) \end{aligned}$$

(c) i. Using the velocity function $v(t) = 9.8(1-t)$, we can find the initial velocity at $t = 0$.

$$v(0) = 9.8(1-0) = 9.8\ m/sec$$

The initial velocity of the ball was $9.8\ m/sec$.

ii. The ball reaches its maximum height when its velocity is zero. Therefore, we can determine the maximum height time as follows.

$$v(t) = 9.8(1-t) = 0$$

The time of maximum height is $t = 1$. This means the ball reached the vertex of the parabola shown in the diagram above at $t = 1$.

iii. We found that the ball reaches its maximum height at $t = 1$. We can calculate the ball's maximum height by plugging this time into the position function.

$$\begin{aligned} s(t) &= 9.8t - 4.9t^2 \\ s(1) &= 9.8(1) - 4.9(1)^2 \\ &= 9.8 - 4.9 \\ &= 4.9\ m \end{aligned}$$

The ball reached a maximum height of 4.9 meters after 1 second.

Part III
Basic Rules
for Differentiation

Chapter 6

Differentiation Techniques

Differentiation is a mathematical technique that allows for calculating derivatives more easily. In this chapter, we will introduce the power rule — one of the fundamental techniques in calculus — for quickly determining the derivative of a function that involves the powers of x. Additionally, we explore the derivative of constant functions, reinforcing key principles that form the foundation of differentiation.

6.1 The Power Rule: An Easy Way to Differentiate

The derivative of the function $f(x) = x^2$ was calculated as $f'(x) = 2x$ through many steps using the derivative's definition we saw in the last chapter. However, finding the derivative will become much more difficult for functions like $f(x) = x^3$ or $f(x) = \sqrt{x}$.

This section will explore a practical differentiation method by observing the repetitive patterns between the functions $f(x)$ and

$f'(x)$. When you look at the graphs of $f(x)$ and $f'(x)$, you'll notice that the derivative function exists in the positive coordinates of the y-axis when the function $f(x)$ increases. Conversely, $f'(x)$ is in the negative coordinates of the y-axis when the function $f(x)$ decreases.

Have you noticed any patterns between the two graphs? The first graph shows a quadratic function for $f(x)$, while $f'(x)$ represents a linear function. Similarly, when $g(x)$ is a cubic function, its derivative, $g'(x)$, is quadratic. Consider the pattern: if $h(x)$ is quartic, $h'(x)$ is cubic. These patterns suggest that **the derivative of a polynomial function always has a degree one less than the original function** before it is differentiated.

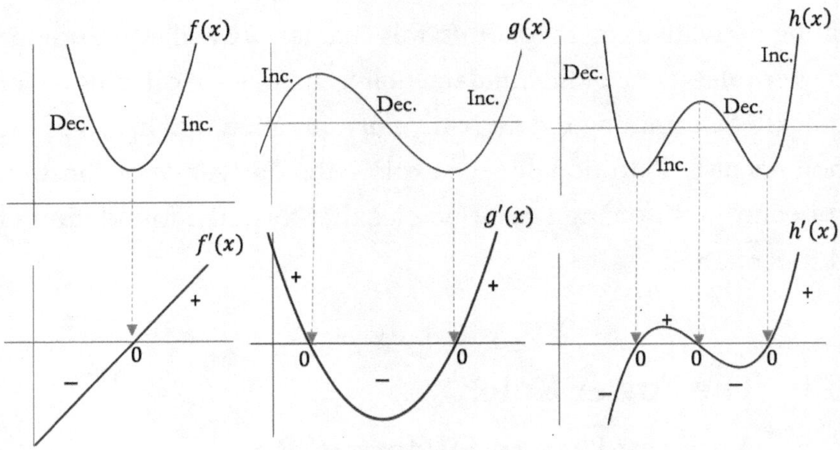

The power rule can be understood graphically by the pattern where the exponent of x decreases by one during differentiation. The **power rule** formula is shown below.

$$f(x) = x^r, \text{ where } r \text{ is a real number}$$

$$f'(x) = rx^{r-1} \text{ or } \frac{d[x^r]}{dx} = rx^{r-1}$$

We can differentiate any exponent using this rule. For example, when we differentiate the quadratic function $f(x) = x^2$, it will be:

$$f'(x) = 2x^{(2-1)} = 2x$$

Similarly, when we differentiate the cubic function $f(x) = x^3$, we discover that $f'(x) = 3x^2$. The differentiations of various functions are displayed in the table.

Function	Differentiation by Power rule	Derivative function
$f(x) = x^3$	$3x^{3-1}$	$f'(x) = 3x^2$
$f(x) = x^{10}$	$10x^{10-1}$	$f'(x) = 10x^9$
$f(x) = \frac{1}{x^2} = x^{-2}$	$-2x^{-2-1}$	$f'(x) = -2x^{-3}$
$f(x) = \sqrt{x} = x^{1/2}$	$\frac{1}{2}x^{\frac{1}{2}-1}$	$f'(x) = \frac{1}{2}x^{-\frac{1}{2}}$
$f(x) = \sqrt{x^3} = x^{3/2}$	$\frac{3}{2}x^{\frac{3}{2}-1}$	$f'(x) = \frac{3}{2}x^{\frac{1}{2}}$
$f(x) = 2x^0 = 2$	$2 \cdot 0x^{0-1}$	$f'(x) = 0$

In the last example, we see the constant function $f(x) = 2$, which is equivalent to $2x^0$, where the variable x has an exponent of 0. Therefore, we can derive another derivative rule, which states that the result of differentiating a constant or constant function is always 0. The reason is that the differentiation of $f(x) = 2 \cdot x^0$ becomes $f'(x) = 2 \cdot 0x^{-1}$ by the power rule. Thus, the result $f'(x)$ becomes 0. The general derivative rule of a constant function can be expressed as:

$$f(x) = c \text{ where c is a constant}$$

$$f'(x) = 0 \ \ or \ \ \frac{d[c]}{dx} = 0$$

The derivative rule for a constant function will be demonstrated graphically. Intuitively, the slope of the constant function, $f(x) = c$, is always 0 at any point along the horizontal line shown below. Therefore, we can conclude that the derivative of a constant function is always 0.

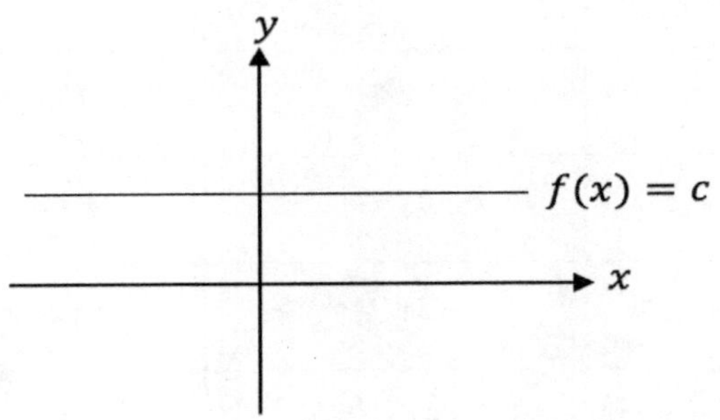

6.2 Differentiating Sums and Differences of Functions

Differentiation is possible even when two functions are connected by addition or subtraction. The differentiation of the sum of two functions, $[f(x) + g(x)]'$, can be determined using the following formula.

$$[f(x) + g(x)]' = f'(x) + g'(x)$$

When two functions are added, the distribution of the prime (′) symbol is possible.

Let us examine how the distribution of prime symbols operates through the process of adding the two functions, as $f(x) = 2x + 3$ and $g(x) = 3x^2 + 2x + 1$, using the following operations. First, we will differentiate each function separately and then combine the results. Second, we will sum the two functions and differentiate the resulting sum. If both results are identical, we can conclude that the distribution of the prime symbol is working correctly. Let us differentiate each function: $f(x) = 2x + 3$ and $g(x) = 3x^2 + 2x + 1$ separately.

$$\begin{aligned} f'(x) &= 2 \cdot 1x^{1-1} + 0 = 2 \\ g'(x) &= 3 \cdot 2x^{2-1} + 2 \cdot 1x^{1-1} + 0 = 6x + 2 \end{aligned}$$

Then, the addition of the results above will be:

$$\begin{aligned} f'(x) + g'(x) &= (2) + (6x + 2) \\ &= 6x + 4 \end{aligned}$$

Now, we will try the other approach of adding the two functions first and then differentiating.

$$\begin{aligned} f(x) + g(x) &= (2x + 3) + (3x^2 + 2x + 1) \\ &= 3x^2 + 4x + 4 \end{aligned}$$

The differentiation of the result above is

$$\begin{aligned} (f(x) + g(x))' &= (3x^2 + 4x + 4)' \\ &= 3 \cdot 2x^{2-1} + 4 \cdot 1x^{1-1} + 0 \\ &= 6x + 4 \end{aligned}$$

The results of the two methods are equivalent. In conclusion, the distribution of the differential sign can be used to solve the differentiation of added functions. Applying the distribution of the differential sign is possible even if the two functions are related by subtraction. We can prove it using the same method as demonstrated above, using the two functions, $f(x) = x^3 + 3x + 1$ and $g(x) = 2x^2 - x$

$$[f(x) - g(x)]' = f'(x) - g'(x)$$

Differentiate each function separately.

$$\begin{aligned} f'(x) &= 3x^{3-2} + 3 \cdot 1x^{1-1} + 0 \\ &= 3x^2 + 3x^0 \\ &= 3x^2 + 3 \\ g'(x) &= 2 \cdot 2x^{2-1} - 1x^{1-1} \\ &= 4x - 1x^0 \\ &= 4x - 1 \end{aligned}$$

The subtraction of the results above will be

$$\begin{aligned} f'(x) - g'(x) &= (3x^2 + 3) - (4x - 1) \\ &= 3x^2 - 4x + 4 \end{aligned}$$

Alternatively, we can first subtract the two functions and then differentiate:

$$\begin{aligned} f(x) - g(x) &= (x^3 + 3x + 1) - (2x^2 - x) \\ &= x^3 - 2x^2 + 4x + 1 \end{aligned}$$

Now, differentiating this result gives:

$$\begin{aligned}[f(x) - g(x)]' &= (x^3 - 2x^2 + 4x + 1)' \\ &= 3x^2 - 4x + 4\end{aligned}$$

From the above demonstration, we can conclude that differentiation rules can be applied to both the sum and the difference of functions. These results are summarized as follows:

$$[f(x) + g(x)]' = f'(x) + g'(x)$$
$$[f(x) - g(x)]' = f'(x) - g'(x)$$

6.3 How to Use the Power Rule in Practice

The power rule can differentiate functions that do not appear as exponent functions, such as $f(x) = \sqrt{x} + \frac{1}{x}$. To apply the power rule, the function must be expressed with an exponent as follows:

$$f(x) = \sqrt{x} + \frac{1}{x} = x^{\frac{1}{2}} + x^{-1}$$

In the process of differentiation using the power rule, each function component undergoes differentiation separately. Then, we can find the result of the derivative of $f(x)$ shown below.

$$
\begin{aligned}
\frac{d}{dx}[f(x)] &= \frac{d}{dx}[x^{\frac{1}{2}} + x^{-1}] \\
&= \frac{d}{dx}[x^{\frac{1}{2}}] + \frac{d}{dx}[x^{-1}] \\
&= \frac{1}{2}x^{\frac{1}{2}-1} + (-1)x^{-1-1} \\
&= \frac{1}{2}x^{-\frac{1}{2}} - x^{-2} \\
&= \frac{1}{2\sqrt{x}} - \frac{1}{x^2}
\end{aligned}
$$

The function $g(x)$ shown below can also be solved with the power rule.

$$g(x) = \frac{1+x+x^2}{x}$$

To apply the power rule to this function, we must express the equation with an exponent using x as a common denominator. Then, each function will be transformed like the one below.

$$\frac{1+x+x^2}{x} = \frac{1}{x} + \frac{x}{x} + \frac{x^2}{x} = x^{-1} + 1 + x$$

We can differentiate each term in the function separately and obtain the $\frac{d}{dx}[g(x)]$ result repeatedly.

$$g(x) = \frac{1+x+x^2}{x} = x^{-1} + 1 + x$$

Then

$$\begin{aligned}
\frac{d}{dx}[g(x)] &= \frac{d}{dx}[x^{-1} + 1 + x] \\
&= \frac{d}{dx}[x^{-1}] + \frac{d}{dx}[1] + \frac{d}{dx}[x] \\
&= -1x^{-1-1} + 0 + 1x^{1-1} \\
&= -x^{-2} + 1x^{0} \\
&= -x^{-2} + 1 \\
&= -\frac{1}{x^{2}} + 1
\end{aligned}$$

6.4 Checkpoint 6

1. Use the power rule to find the derivatives of the following functions.

   (a) $f(x) = x^2 + x + 1$

   (b) $f(x) = 2x^5 - 3x^3 + 7x$

   (c) $f(x) = -5x^{-5} + 3x^{-2} - 7x^{-1}$

2. Find the derivative function using the power rule.

   (a) $f(x) = \frac{1}{x^2} + \frac{2}{x^3} + \frac{3}{x^4}$

      i. Rewrite the given function using a negative exponent. Hint: $\frac{1}{x^n} = x^{-n}$

      ii. Find $f'(x)$.

   (b) $f(x) = \sqrt{x} + \sqrt[3]{x} + \sqrt[4]{x}$

      i. Rewrite the radical function using the rational exponent. Hint: $\sqrt[n]{x} = x^{\frac{1}{n}}$

      ii. Find $f'(x)$.

3. After converting the function to include an exponent using the common denominator, find $f'(x)$.

   (a) $f(x) = \frac{x^2+x^4}{x}$

   (b) $f(x) = \frac{1+x^2+x^4}{x^5}$

4. Given two functions $f(x) = 5x^2 + 1$ and $g(x) = 4x^3 + 2x^2 + x$.

   (a) Find the derivatives of $f(x)$ and $g(x)$ separately, then add the results together.

(b) Find the sum of $f(x)$ and $g(x)$, and differentiate the result. After that, prove the following formula.

$$[f(x)+g(x)]' = f'(x)+g'(x)$$

(c) Find the derivatives of $f(x)$ and $g(x)$ separately, then subtract the results.

(d) Find the difference of $f(x)$ and $g(x)$, and differentiate the results. After that, prove the following formula.

$$[f(x)-g(x)]' = f'(x)-g'(x)$$

6.5 Checkpoint 6 Solution

1. (a)

$$\begin{aligned} f'(x) &= 2x + 1 + 0 \\ &= 2x + 1 \end{aligned}$$

(b)

$$\begin{aligned} f'(x) &= 2 \cdot 5x^4 - 3 \cdot 3x^2 + 7 \cdot 1x^0 \\ &= 10x^4 - 9x^2 + 7 \end{aligned}$$

(c)

$$\begin{aligned} f'(x) &= -5 \cdot (-5)x^{-6} + 3(-2)x^{-3} - 7(-1)x^{-2} \\ &= 25x^{-6} - 6x^{-3} + 7x^{-2} \end{aligned}$$

2. (a) i.

$$\frac{1}{x^2} + \frac{2}{x^3} + \frac{3}{x^4} = x^{-2} + 2x^{-3} + 3x^{-4}$$

ii.

$$\begin{aligned} f'(x) &= -2x^{-3} + 2(-3)x^{-4} + 3(-4)x^{-5} \\ &= -2x^{-3} - 6x^{-4} - 12x^{-5} \\ &= -\frac{2}{x^3} - \frac{6}{x^4} - \frac{12}{x^5} \end{aligned}$$

(b) i.

$$\sqrt{x}+\sqrt[3]{x}+\sqrt[4]{x}=x^{\frac{1}{2}}+x^{\frac{1}{3}}+x^{\frac{1}{4}}$$

ii.

$$f'(x)=\frac{1}{2}x^{-\frac{1}{2}}+\frac{1}{3}x^{-\frac{2}{3}}+\frac{1}{4}x^{-\frac{3}{4}}$$

3. (a)

$$\begin{aligned}f(x)&=\frac{x^2+x^4}{x}\\&=\frac{x^2}{x}+\frac{x^4}{x}\\&=x+x^3\\f'(x)&=1+3x^2\end{aligned}$$

(b)

$$\begin{aligned}f(x)&=\frac{1+x^2+x^4}{x^5}\\&=\frac{1}{x^5}+\frac{x^2}{x^5}+\frac{x^4}{x^5}\\&=x^{-5}+x^{-3}+x^{-1}\\f'(x)&=-5x^{-6}-3x^{-4}-x^{-2}\\&=-\frac{5}{x^6}-\frac{3}{x^4}-\frac{1}{x^2}\end{aligned}$$

4. (a)

$$\begin{aligned} f'(x) &= 10x \\ g'(x) &= 12x^2 + 4x + 1 \\ f'(x) + g'(x) &= (10x) + (12x^2 + 4x + 1) \\ &= 12x^2 + 14x + 1 \end{aligned}$$

(b)

$$\begin{aligned} f(x) + g(x) &= (5x^2 + 1) + (4x^3 + 2x^2 + x) \\ &= 4x^3 + 7x^2 + x + 1 \\ [f(x) + g(x)]' &= 12x^2 + 14x + 1 \end{aligned}$$

The results of $[f(x) + g(x)]'$ and $f'(x) + g'(x)$ are the same.

(c)

$$\begin{aligned} f'(x) &= 10x \text{ and } g'(x) = 12x^2 + 4x + 1 \\ f'(x) - g'(x) &= (10x) - (12x^2 + 4x + 1) \\ &= -12x^2 + 6x - 1 \end{aligned}$$

(d)

$$\begin{aligned} f(x) - g(x) &= (5x^2 + 1) - (4x^3 + 2x^2 + x) \\ &= -4x^3 + 3x^2 - x + 1 \\ [f(x) - g(x)]' &= -12x^2 + 6x - 1 \end{aligned}$$

The results of $[f(x) - g(x)]'$ and $f'(x) - g'(x)$ are the same.

Chapter 7

Product and Quotient Rules

7.1 Product Rule

We have learned that differential symbols can be distributed over the sum and difference of functions. However, can the same principle be applied to the product of two functions? In other words, does the derivative of the product, $[f(x) \cdot g(x)]'$ equal the product of the derivative, $f'(x) \cdot g'(x)$?. Let's examine the functions $f(x) = x$ and $g(x) = x^2$. Before proceeding with the proof, we first summarize the necessary information.

$$f(x) = x \quad \text{and} \quad g(x) = x^2$$

$$f(x) \cdot g(x) = x \cdot x^2 = x^3$$

After multiplying the two functions, we can obtain the result $[f(x) \cdot g(x)]'$ through differentiation.

$$[f(x) \cdot g(x)]' = (x^3)' = 3x^2$$

However, the results of multiplying each function's derivative differ, as shown below:

$$f'(x) \cdot g'(x) = 1 \cdot 2x = 2x$$

Therefore, we can conclude that the results for $[f(x) \cdot g(x)]'$ and $f'(x) \cdot g'(x)$ are different since

$$3x^2 \neq 2x$$

It is important to note that the distributive property does not apply to the multiplication of functions.

$$[f(x) \cdot g(x)]' \neq f'(x) \cdot g'(x)$$

Our exploration shows that the derivative of a product is not simply the product of the derivatives. This means we need a new rule for differentiation: the **product rule**.

$$[f(x) \cdot g(x)]' = f(x) \cdot g'(x) + g(x) \cdot f'(x)$$

Upon applying the rule, we can find the derivative of the product of $f(x) = x$ and $g(x) = x^2$:

$$f(x) = x \quad \text{and} \quad f'(x) = 1$$

$$g(x) = x^2 \quad \text{and} \quad g'(x) = 2x$$

Therefore, $[f(x) \cdot g(x)]'$ is

$$\begin{aligned} f(x) \cdot g'(x) + g(x) \cdot f'(x) &= (x) \cdot (2x) + (x^2) \cdot (1) \\ &= 2x^2 + x^2 \\ &= 3x^2 \end{aligned}$$

The product rule is useful when differentiating functions that involve multiplication, such as:

$$h(x) = 2x^3(x^5 - 7x)$$

There are two ways to find the derivative of $h(x)$. One way is to expand the expression and then apply the power rule. The other way is to use the product rule. Start by expanding $h(x)$ and rearranging the terms to see what we have:

$$\begin{aligned} h(x) &= 2x^3(x^5 - 7x) \\ &= 2x^3 \cdot x^5 - 2x^3 \cdot 7x \\ &= 2x^8 - 14x^4 \end{aligned}$$

Afterwards, we can find $h'(x)$ from $h(x)$ using the power rule:

$$\begin{aligned} h'(x) &= 2 \cdot 8x^7 - 14 \cdot 4x^3 \\ &= 16x^7 - 56x^3 \end{aligned}$$

Alternatively, by applying the product rule, we can differentiate $f(x) \cdot g(x)$ and have the same result.

$$f(x) = 2x^3 \quad \text{and} \quad g(x) = x^5 - 7x$$

$$f'(x) = 6x^2 \quad \text{and} \quad g'(x) = 5x^4 - 7$$

Then,

$$\begin{aligned}[f(x) \cdot g(x)]' &= f(x) \cdot g'(x) + g(x) \cdot f'(x) \\ &= (2x^3) \cdot (5x^4 - 7) + (x^5 - 7x) \cdot (6x^2) \\ &= 10x^7 - 14x^3 + 6x^7 - 42x^3 \\ &= 16x^7 - 56x^3\end{aligned}$$

We can see that differentiations give the same result whether we apply the product rule directly or expand the expression first and then use the power rule. As functions become more complex, applying the product rule is often the more practical and efficient approach.

7.2 Quotient Rule

Similar to multiplication, differentiation cannot distribute over division:

$$\left[\frac{f(x)}{g(x)}\right]' \neq \frac{f'(x)}{g'(x)}$$

Instead, we can use the **quotient rule** below. Like the product rule, it is also recommended that you memorize the following formula:

$$\left[\frac{f(x)}{g(x)}\right]' = \frac{f'(x) \cdot g(x) - f(x) \cdot g'(x)}{[g(x)]^2}$$

We will practice the derivative of the quotient function using the values of $f(x)$ and $g(x)$ given below.

$$f(x) = x^3 + x \text{ and } g(x) = x^2 + 2x$$

Let's define

$$h(x) = \frac{f(x)}{g(x)}$$

then the $h'(x)$ will be as follows:

$$\begin{aligned} h'(x) &= \left[\frac{x^3 + x}{x^2 + 2x}\right]' \\ &= \frac{(3x^2+1)\cdot(x^2+2x) - (x^3+x)\cdot(2x+2)}{(x^2+2x)^2} \\ &= \frac{3x^4 + x^2 + 6x^3 + 2x - 2x^4 - 2x^3 - 2x^2 - 2x}{x^4 + 4x^3 + 4x^2} \\ &= \frac{x^4 + 4x^3 - x^2}{x^4 + 4x^3 + 4x^2} \\ &= \frac{x^2(x^2 + 4x - 1)}{x^2(x^2 + 4x + 4)} \\ &= \frac{x^2 + 4x - 1}{x^2 + 4x + 4} \end{aligned}$$

In conclusion, when two functions are related by multiplication and division, the product rule or quotient rule should be used instead of directly distributing the differential sign.

Summary of Techniques for Differentiation

- **Derivative of a constant:**

$$\frac{d}{dx}[c] = 0$$

- **Power rule:**

$$\frac{d}{dx}[x^n] = nx^{n-1}$$

- **Sum and difference rule:**

$$(f+g)' = f' + g'$$

$$(f-g)' = f' - g'$$

- **Product and quotient rule:**

$$(f \cdot g)' = f \cdot g' + g \cdot f'$$

$$\left(\frac{f}{g}\right)' = \frac{f' \cdot g - f \cdot g'}{g^2}$$

7.3 Checkpoint 7

1. Determine the derivative functions of the following.

   (a) $f(x) = (2x+1)(x^2-1)$

   (b) $g(x) = (5x^3+2x)(3x^2-1)$

2. The derivatives of some trigonometric functions are

$$f(x) = \sin x,\ f'(x) = \cos x$$

$$g(x) = \cos x,\ g'(x) = -\sin x$$

   Find the derivative of $h(x) = \sin x \cdot \cos x$ using the product rule.

3. Determine the derivative functions of the following problems using the quotient rule.

   (a) $f(x) = \frac{x^2+1}{2x-5}$

   (b) $g(x) = \frac{\sin x}{\cos x}$

4. Adam owns a small coffee stand in the park. Through observation, he notices a relationship between price and quantity sold; fewer units are sold as the price increases. When the coffee is free (price = \$0), he sells 500 cups. For every \$1 increase in price, sales decrease by 50 cups. From this information, he models the quantity sold $Q(p)$ as dependent on price:

$$Q(p) = 500 - 50p$$

In addition, he derives the total revenue $R(p)$ as follows:

$$R(p) = p \cdot Q(p) = p \cdot (500 - 50p)$$

(a) Using the product rule, find the derivative of the revenue function $R'(p)$.

(b) Explain the meaning of $R'(p)$ in this context. What does the derivative of revenue with respect to price represent?

(c) At what price is revenue maximized? Use your derivative to find the stationary point.

7.4 Checkpoint 7 Solution

1. (a) $(2x+1)' = 2$ and $(x^2-1)' = 2x$,

$$\begin{aligned} f'(x) &= (2x+1)\cdot(x^2-1)' + (x^2-1)\cdot(2x+1)' \\ &= (2x+1)(2x) + (x^2-1)(2) \\ &= 4x^2+2x+2x^2-2 \\ &= 6x^2+2x-2 \end{aligned}$$

   (b) $(5x^3+2x)' = 15x^2+2$ and $(3x^2-1)' = 6x$,

$$\begin{aligned} g'(x) &= (5x^3+2x)\cdot(3x^2-1)' + (3x^2-1)\cdot(5x^3+2x)' \\ &= (5x^3+2x)(6x) + (3x^2-1)(15x^2+2) \\ &= 30x^4+12x^2+45x^4+6x^2-15x^2-2 \\ &= 75x^4+3x^2-2 \end{aligned}$$

2.

$$\begin{aligned} h' &= f(x)\cdot g'(x) + g(x)\cdot f'(x) \\ &= (\sin x)\cdot(-\sin x) + (\cos x)\cdot(\cos x) \\ &= -\sin^2 x + \cos^2 x \\ &= \cos^2 x - \sin^2 x \end{aligned}$$

3. (a) $(x^2+1)' = 2x$ and $(2x-5)' = 2$,

$$\begin{aligned} f'(x) &= \frac{(2x-5)\cdot(2x)-(x^2+1)\cdot(2)}{(2x-5)^2} \\ &= \frac{4x^2-10x-2x^2-2}{(2x-5)^2} \\ &= \frac{2x^2-10x-2}{(2x-5)^2} \end{aligned}$$

(b) $(\sin x)' = \cos x$ and $(\cos x)' = -\sin x$,

$$\begin{aligned} g'(x) &= \frac{(\cos x)\cdot(\cos x)-(\sin x)\cdot(-\sin x)}{\cos^2 x} \\ &= \frac{\cos^2 x+\sin^2 x}{\cos^2 x} \\ &= \frac{\cos^2 x}{\cos^2 x}+\frac{\sin^2 x}{\cos^2 x} \\ &= 1+\frac{\sin^2 x}{\cos^2 x} \end{aligned}$$

4. (a) $R(p) = p\cdot(500-50p)$. Let $f(p) = p$ and $g(p) = (500-50p)$. Then $R'(p)$ can be derived as follows by the product rule:

$$\begin{aligned} R'(p) &= p'\cdot(500-50p)+p\cdot(500-50p)' \\ &= (1)(500-50p)+p(-50) \\ &= 500-50p-50p \\ &= 500-100p \end{aligned}$$

(b) The derivative of the revenue function, $R'(p) = 500 - 100p$, indicates the rate at which revenue changes with respect to price. This equation is essential for understanding how revenue varies with price adjustments.

- Positive slope ($R'(p) > 0$): When the derivative is positive, revenue increases as the price increases.
- Negative slope ($R'(p) < 0$): When the derivative is negative, revenue decreases as the price increases.
- Zero Derivative ($R'(p) = 0$): A zero derivative means the revenue is at a turning point (either a maximum or a minimum). Since the graph of the revenue function opens downward, this turning point corresponds to the maximum revenue.

Understanding these principles is essential for effective pricing strategies.

(c) To find the price that maximizes revenue, we must determine the value of p for which the derivative is equal to zero.

$$\begin{aligned} R'(p) &= 0 \\ 500 - 100p &= 0 \\ 500 &= 100p \\ p &= 5 \end{aligned}$$

Therefore, Adam maximizes his revenue when the price is \$5. By substituting this price as $p = 5$ into both functions, $Q(p)$ and $R(p)$, we can find the amount of coffee sold and the maximum revenue as follows:

$$Q(5) = 500 - 50(5) = 250 \text{ cups}$$

$$R(5) = 5 \cdot 250 = \$1250$$

In conclusion, the maximum revenue is $1250 when the coffee price is $5 and the coffee sold is 250 cups.

Chapter 8

Chain Rule

Have you ever heard the phrase 'a good influence spreads'? A small act of kindness can move or affect others unexpectedly. Interestingly, a similar idea exists in mathematics! When one function feeds into another, its influence affects the overall outcome.

This concept is known as function composition, where the output of one function becomes the input of another. Function composition appears throughout mathematics, making it essential to understand how to differentiate these combined functions. In this chapter, we'll explore the **chain rule**, a powerful tool for finding the derivative of composite functions.

8.1 Understanding and Applying the Chain Rule

Continuing with the idea of one thing influencing another, let's consider a real-life example. Imagine your family rents a car for a trip. The rental employee explains that 50 liters of fuel costs \$80 and will let the car travel 400 km. This means the amount you spend on fuel directly affects how far you can travel: more liters of fuel cost more money and also take you farther. In this example, liters act as the bridge between dollars and distance — change the liters, and both the cost and the travel range change.

This question becomes much easier if you break it down step by step. By knowing how far the car can travel per dollar spent, you can connect the relationships between distance and fuel, as well as fuel and cost. Just like in math, where one function influences another, understanding these relationships helps us make calculations efficiently as follows:

$$\frac{400\ km}{50\ L} \times \frac{50\ L}{\$80} = \frac{5\ km}{1\$}$$

As calculated above, you can drive 5 km for each dollar. In this calculation, it is clear that two functions are linked. The first function shows the relationship between the distance a car travels and its fuel consumption. The second function illustrates the relationship between fuel consumption and cost. By connecting these two functions, we demonstrate the correlation between travel distance and cost. This is called function composition, since one function is used as input for the other and influences the output, as shown below.

$$\frac{\text{travel distance}}{\text{fuel consumption}} \times \frac{\text{fuel consumption}}{\text{cost}} = \frac{\text{travel distance}}{\text{cost}}$$

In mathematics, a function represents the relationship between input and output values. In the previous example, as we connected fuel consumption, cost, and travel distance, we can visualize how the output of one function feeds into another. The diagram below illustrates function composition, where the output of the first function becomes the input for the next.

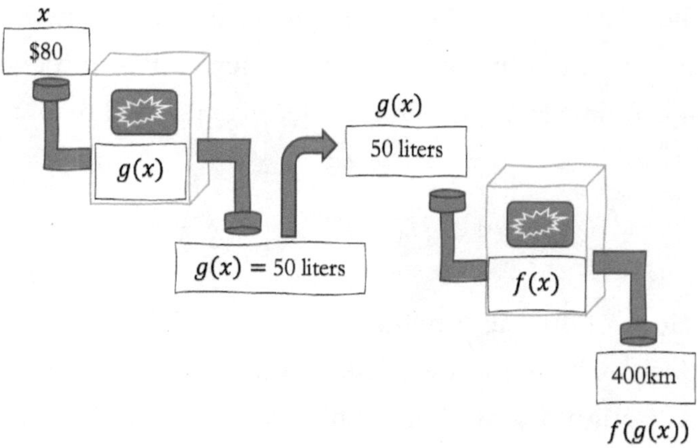

The composition function is represented by $f(g(x))$, where the output value of the function $g(x)$ is used as the input for the function $f(x)$. For example, the function $g(x)$ calculates the amount of fuel when the cost is entered. When the input value is entered as \$80 into $g(x)$, the machine's output will be 50 liters. The function $f(x)$ converts the amount of fuel to the travel distance. Therefore, 50 liters go into the input value, and the output will be 400km. By connecting these functions, $f(x)$ and $g(x)$, we can get a new function, $f(g(x))$, which represents the car's mileage according to the cost.

The following two functions are given: $f(x) = x^3$ and $g(x) = x^2 + 2$. When the function $g(x) = x^2 + 2$ is entered into the input of the function $f(x) = x^3$, $f(x)$ becomes $f(g(x))$, and the result is $(x^2 + 2)^3$ as demonstrated below.

$$f(x) = x^3 \text{ and } g(x) = x^2 + 2$$

$$f(g(x)) = f(x^2 + 2) = (x^2 + 2)^3$$

When you look at the result of the composition function, $(x^2 + 2)^3$, the outer function $(\ )^3$ wraps the inner function x^2+2. In the composite function $f(g(x))$, $f(x)$ serves as the outer function, while $g(x)$ represents the inner function. A derivative of a composite function is found using the chain rule, as explained below.

$$\frac{d}{dx}[f(g(x))] = [f(g(x))]' = f'(g(x)) \cdot g'(x)$$

This idea of linking functions naturally leads to the chain rule, which derives its name from the chain-like process of differentiation. Although it may seem complex at first, the rule follows a straightforward pattern: **first, differentiate the outer function, and multiply it by the derivative of the inner function.**

To see this in action, let's differentiate the composite function $f(g(x)) = (x^2+2)^3$ step by step. In this function, the outer function is $(\ )^3$, and the inner function is x^2+2. First, we obtain the following result by applying the power rule to the outer function.

$$\begin{aligned}\frac{d\left[(x^2+2)^3\right]}{dx} &= 3(x^2+2)^{3-1} \\ &= 3(x^2+2)^2\end{aligned}$$

When finding the derivative of a composite function, start by differentiating the outer function while keeping the inner function unchanged inside the parentheses ( ). You don't need to worry about the inner function yet. Once the outer function is done, differentiate the inner function.

$$(x^2+2)' = 2x + 0$$

Now, multiply the above two results to complete the chain rule.

$$\begin{aligned}[(x^2+2)^3]' &= 3(x^2+2)^2 \cdot 2x \\ [f(g(x))]' &= f'(g(x)) \cdot g'(x) \\ [\text{Composite function}]' &= \text{Outer function}' \cdot \text{ Inner function}'\end{aligned}$$

The result of the differentiation of the composite function can be summarized as follows.

$$\begin{aligned}[f(g(x))]' &= [(x^2+2)^3]' \\ &= 3(x^2+2)^2 \cdot 2x \\ &= 6x(x^2+2)^2\end{aligned}$$

To apply the chain rule, we must identify the outer function and the inner function before differentiating a composite function. For further practice, try finding the derivative of a composite function using another example.

$$y = \sqrt{x^2 - 2x + 3}$$

What is the outer function in the example above? If it's not clear, try rewriting it as an exponent, like this:

$$y = \sqrt{x^2 - 2x + 3} = (x^2 - 2x + 3)^{\frac{1}{2}}$$

Have you found it? The outer function is $(\ )^{\frac{1}{2}}$ and the inner function is $x^2 - 2x + 3$. The differentiation of the outer function will be

$$\left[(\ )^{\frac{1}{2}}\right]' = \frac{1}{2}(\ )^{\frac{1}{2}-1}$$

The differentiation of the inner function will be

$$(x^2 - 2x + 3)' = 2x - 2 + 0$$

Using the chain rule to differentiate this function, the result will be as follows.

$$\begin{aligned}\left(\sqrt{x^2 - 2x + 3}\right)' &= \frac{1}{2}(x^2 - 2x + 3)^{-1/2} \cdot (2x - 2) \\ &= \frac{2x - 2}{2\sqrt{x^2 - 2x + 3}} \\ &= \frac{x - 1}{\sqrt{x^2 - 2x + 3}}\end{aligned}$$

8.2 Using Substitution with the Chain Rule

An alternative to applying the chain rule directly is to substitute the inner function $g(x)$ with a new variable u. This transforms the composite function $f(g(x))$ into $f(u)$, making the differentiation easier.

For example, consider:

$$(x^2 + 1)^{57} = u^{57}$$

where, $u = x^2 + 1$ and $u' = 2x$. The following process shows the result of differentiation.

$$\begin{aligned}\left(u^{57}\right)' &= 57u^{56} \cdot u' \\ &= 57(x^2 + 1)^{56} \cdot 2x \\ &= 114x(x^2 + 1)^{56}\end{aligned}$$

The general form of the chain rule can be expressed as:

$$[f(g(x))]' = f'(g(x)) \cdot g'(x)$$

or equivalently, using substitution,

$$[f(u)]' = f'(u) \cdot u'$$

The chain rule is one of the most important techniques in calculus and is widely used when differentiating composite functions. Let's practice more in the checkpoint.

8.3 Checkpoint 8

1. Using the chain rule below:

$$[f(g(x))]' = f'(g(x)) \cdot g'(x)$$

   Find the derivative function of the following.

   (a) $y = (x^2 - 5)^3$

      i. What are the outer and the inner functions?

      ii. What is the derivative function, y?

   (b) $y = (3x^2 + 5x - 1)^{37}$

      i. What are the outer and the inner functions?

      ii. What is the derivative function, y?

2. Find the derivative y of the function $y = \sqrt{x^3 - 2x}$.

3. Using the given values, $f'(2) = 5$, $g(3) = 2$, and $g'(3) = 1$, find the value of $[f(g(3))]'$.

4. Using the given values, $f'(5) = 7, g(0) = 5$, and $g'(0) = -2$, find the following value.

5. Using the following values, $f(2) = 3$ and $f'(2) = 7$, find the result of $h'(2)$ using the following function.

$$h(x) = [f(x)]^2$$

8.4 Checkpoint 8 Solution

1. (a) For $y = (x^2 - 5)^3$:

$$\begin{aligned} \text{Outer function: } & f(u) = u^3 \\ \text{Inner function: } & g(x) = x^2 - 5 \\ y' &= f'(g(x)) \cdot g'(x) \\ &= 3(x^2-5)^2 \cdot 2x \\ &= 6x(x^2-5)^2 \end{aligned}$$

   (b) For $y = (3x^2 + 5x - 1)^{37}$:

$$\begin{aligned} \text{Outer function: } & f(u) = u^{37} \\ \text{Inner function: } & g(x) = 3x^2 + 5x - 1 \\ y' &= f'(g(x)) \cdot g'(x) \\ &= 37(3x^2+5x-1)^{36} \cdot (6x+5) \end{aligned}$$

2. For $y = \sqrt{x^3 - 2x}$:

$$\begin{aligned} \text{Outer function: } & f(u) = u^{1/2} \\ \text{Inner function: } & g(x) = x^3 - 2x \\ y' &= f'(g(x)) \cdot g'(x) \\ &= \frac{1}{2}(x^3-2x)^{-1/2} \cdot (3x^2-2) \end{aligned}$$

3. For $[f(g(3))]'$ with given values:

$$
\begin{aligned}
f'(2) = 5, \quad g(3) = 2, \quad g'(3) = 1 \\
[f(g(3))]' &= f'(g(3)) \cdot g'(3) \\
&= f'(2) \cdot g'(3) \\
&= (5) \cdot (1) = 5
\end{aligned}
$$

4. For $[f(g(0))]'$ with given values:

$$
\begin{aligned}
f'(5) = 7, \quad g(0) = 5, \quad g'(0) = -2 \\
[f(g(0))]' &= f'(g(0)) \cdot g'(0) \\
&= f'(5) \cdot g'(0) \\
&= (7) \cdot (-2) = -14
\end{aligned}
$$

5. For $h(x) = [f(x)]^2$ with given values:

$$
\begin{aligned}
f(2) = 3, \quad f'(2) = 7 \\
h'(x) = 2f(x) \cdot f'(x) \\
h'(2) = 2 \cdot (3) \cdot (7) = 42
\end{aligned}
$$

Part IV
Applying Derivatives to Real-World Problems

Chapter 9

Understanding Change in Real-World Contexts

Calculus is not just a subject in textbooks — it's something we use every day without even realizing it. Whenever we think about how fast something is changing, whether it's the price of a product, the speed of a car, or minimizing costs to grow a business, we use the ideas behind calculus. In particular, the derivatives in calculus provide a method to measure and predict these changes.

In this chapter, we explore real-life examples through three different scenarios. The first illustrates calculus as a powerful problem-solving tool. The second and third involve making informed decisions, such as analyzing motion or optimizing resources.

9.1 Rate of Change in Business Applications

Scenario 1: Adam's Iced Coffee Stand

Adam owns a small iced coffee stand in the park. He noticed that when the coffee was free, he could sell 500 cups. As he raises the price, fewer people buy the coffee; for every \$1 increase in price, he sells 50 fewer units. Through experience, Adam formulated a simple equation as follows:

1. When the price (p) is zero (free), the quantity sold (q) amounts to 500 units. This signifies the starting point or the maximum number of units that can be sold, although the revenue will still be zero.

$$q = 500 - 50(0)$$

2. He notes that the decrease in demand corresponds to a reduction of 50 units in sales for every \$1 increase in price. Thus, if the price is represented by p, the decline in sales can be expressed as $50p$.

$$q = 500 - 50p$$

3. He indicates that the quantity of sales depends on the price, and quality is a function of price. Hence, he derives the final equation for the total quantity sold as the starting amount minus the decrease:

$$q = f(p) = 500 - 50p$$

The equation above shows that if the price is \$5 ($p = 5$), the quantity sold is:

$$f(5) = 500 - 50(5) = 500 - 250 = 250$$

This means that 250 units are sold when the price is \$5. Similarly, if the price increases to \$10 ($p = 10$), the quantity sold is:

$$f(5) = 500 - 50(10) = 500 - 500 = 0$$

This result indicates that for $10, no units are sold — the demand drops to zero.

Using the function above, $f(p) = 500 - 50p$, Adam aims to find the optimal price to maximize his revenue. To understand this process, let's carefully examine what derivatives mean and how they help us identify the maximum.

This derivative, $f'(p)$ *or* $\frac{dq}{dp}$, shows how the quantity demanded responds to a change in price. In terms of units, $f(p)$ is measured in cups of iced coffee, while p is measured in dollars. Therefore, $f'(p)$ has a unit of cups per dollar — it indicates how many cups of iced coffee sales change when the price shifts by one dollar.

To describe the explanation above mathematically, we need to differentiate the function $f(p)$.

$$\begin{aligned} f(p) &= 500 - 50p \\ f'(p) &= 0 - 50 \\ &= -50 \end{aligned}$$

The result, $f'(p) = -50 \frac{cups}{\$}$, describes that for every $1 increase in price, Adam sells 50 fewer cups per day.

For example, if Adam is currently selling iced coffee at $4 per cup, the value $f'(p) = -50$, this indicates that for every $1 increase in price, Adam will sell 50 fewer cups. Conversely, for every $1 decrease, he will sell 50 more cups. Therefore, raising the price from $4 to $5 would decrease sales by about 50 cups, while lowering the price from $4 to $3 would increase sales by about 50 cups.

Based on the information above, Adam sees that increasing the price decreases the number of cups sold, while lowering the price increases sales. Understanding this relationship, Adam wants to make a smart decision to maximize his revenue using the following revenue formula:

$$\text{Revenue} = \text{Price per unit} \times \text{Quantity sold}$$

Using this, Adam formulates the revenue equation as follows:

$$R = p \times f(p)$$

Substituting, $f(p) = 500 - 50p$, we can make a revenue function which depends on the price as follows:

$$R(p) = p \times (500 - 50p)$$

To find the price that maximizes revenue, we are looking for the highest point of the revenue function $R(p)$. Imagine climbing a hill: as you go up, the slope is positive; as you go down, the slope is negative. Exactly at the very top — the peak — the flattening of the hill is zero. Similarly, the maximum of a function occurs where its derivative equals zero. Thus, to maximize revenue, we calculate the derivative of $R(p)$ and set it equal to zero.

The revenue function is given by:

$$R(p) = p \times (500 - 50p)$$

Expanding the expression:

$$R(p) = 500p - 50p^2$$

Differentiate $R(p)$ by the power rule with respect to p:

$$R'(p) = 500 - 100p$$

To find the price that maximizes revenue, set $R'(p) = 0$:

$$500 - 100p = 0$$

and we can solve for the price p:

$$100p = 500$$

$$p = 5$$

Next, substitute $p = 5$ into the revenue function to find the maximum revenue:

$$R(5) = 5(500 - 50(5)) = 1250$$

We can check if $p = 5$ truly gives the highest revenue by comparing it to nearby values, such as $p = 4$ and $p = 6$:

$$R(4) = 4(500 - 50(4)) = 1200$$

$$R(6) = 6(500 - 50(6)) = 1200$$

The graph below also illustrates the relationship between price and revenue, confirming that a price of \$5 results in the maximum revenue of \$1250.

This scenario is a good example to see how calculus applies to business and economics as well as how it helps in making smart decisions, such as maximizing profits. In this example, we learn how changing the price of his coffee affects the number of cups sold and, in turn, the total revenue. It also uses the concept of derivatives, which helps us understand how one quantity changes in response to another.

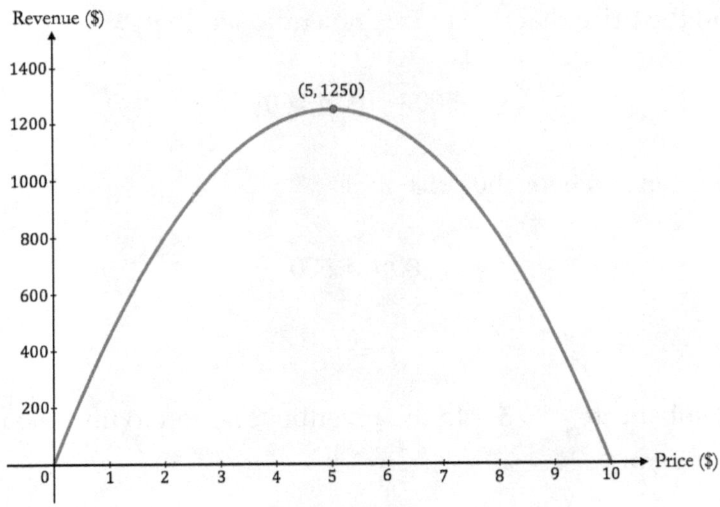

By analyzing the rate of change of sales with respect to price, we can determine the optimal price to maximize revenue. Through Adam's story, you'll see how calculus is not just about numbers but also about making better choices in everyday life.

9.2 Understanding the Motion of Two Moving Objects

Scenario 2: A Related Rates Bicycle and Skateboard

Adam is riding his bicycle north toward an intersection, while his friend Hannah is skateboarding east, having already passed through it as shown in the diagram below. At a particular moment, Adam is 0.3 miles south of the intersection and is biking at 10 mph toward it. Hannah is 0.4 miles east of the intersection and is skateboarding at an unknown speed. A GPS tracker shows that the distance between Adam and Hannah is increasing at 4 mph.

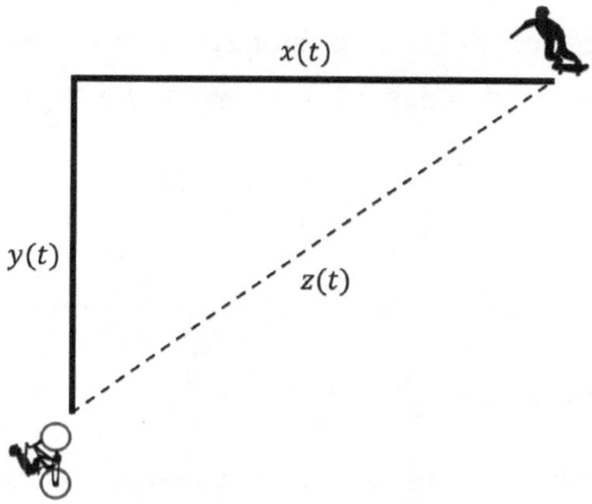

Using the information above, we can find Hannah's skateboarding speed. To answer this question, we first need to define the variable as follows:

- $x(t)$: The distance of Hannah from the intersection (moving east).
- $y(t)$: The distance of Adam from the intersection (moving north).
- $z(t)$: The distance between Hannah and Adam.

Since Adam and Hannah form a right triangle with the intersection, we can use the Pythagorean theorem to relate their distances:

$$z^2 = x^2 + y^2$$

It's important for us to take the derivative with respect to t (time) instead of any other variable, since all the variables involved $x, y,$

and z change over time. This means that time (t) is the independent variable that determines the values of x, y, and z at any moment.

Differentiating both sides with respect to t, we get:

$$2z\frac{dz}{dt} = 2x\frac{dx}{dt} + 2y\frac{dy}{dt}$$

At this moment, we can summarize the given values as follows:

- $x = 0.4$ mile and $y = 0.3$ mile
- $\frac{dy}{dt} = -10$ mph (Adam's cycling speed and the sign is negative since he is moving toward the intersection)
- $\frac{dz}{dt} = 4$ mph (The rate of distance change between Adam and Hannah)

We need to find z in the moment to solve this question using the Pythagorean theorem.

$$z = \sqrt{x^2 + y^2} = \sqrt{0.4^2 + 0.3^2} = 0.5$$

Now, substituting values into the differentiated equation to determine Hannah's speed $\frac{dx}{dt}$:

$$\begin{aligned}
2z\frac{dz}{dt} &= 2x\frac{dx}{dt} + 2y\frac{dy}{dt} \\
2(0.5)(4) &= 2(0.4)\frac{dx}{dt} + 2(0.3)(-10) \\
4 &= 0.8\frac{dx}{dt} - 6 \\
\frac{dx}{dt} &= \frac{10}{0.8} \\
\frac{dx}{dt} &= 12.5
\end{aligned}$$

Currently, Hannah is travelling 12.5 mph away from the intersection. This example shows how calculus can analyze motion in real-life situations, like how the distance between two moving objects changes over time.

This scenario is an example of related rates, a key concept in calculus that helps us understand how one quantity changes in relation to another over time. By analyzing this scenario, we learn how to use derivatives to determine the rate at which the distance between two moving objects is changing. This concept is essential not only in physics and engineering but also in everyday applications, such as tracking the movement of objects, predicting changes in speed, and solving real-time problems in navigation and surveillance.

9.3 Optimization

Scenario 3: Minimizing Material Costs

Adam, an engineer at an oil company, is designing a one-liter oil can in the shape of a right circular cylinder. His goal is to minimize the amount of material used for the can, which means he wants to find the dimensions (radius and height) that require the least surface area while still holding exactly one liter of oil.

For performing this project, Adam defined the variables needed as follows:

- r be the radius of the base (in cm).
- h be the height of the cylinder (in cm).
- The volume of the cylinder is 1 liter or 1000 cm^3 (since 1 liter = 1000 cm^3).

The formula for the volume of a cylinder is:

$$V = \pi \cdot r^2 \cdot h$$

Since we need the volume to be 1000 cm³, we set up the equation:

$$1000 = \pi \cdot r^2 \cdot h$$

The surface area of a can includes:

- The top and bottom circles: $2 \times (\pi \cdot r^2)$
- The curved lateral surface area: $2 \cdot \pi \cdot r \cdot h$

So, the total surface area (A) of the oil can is:

$$A = 2\pi r^2 + 2\pi rh$$

The height (h) appears in both the volume equation and the surface area equation. To simplify the calculation, we would better reduce the number of variables by expressing the surface area in terms of a single variable. This process is important because differentiation is much easier when dealing with one variable instead of multiple variables. Since our goal is to minimize the surface area while keeping the volume fixed, we use the volume equation to eliminate h and rewrite the surface area in terms of a single variable as follows:

$$h = \frac{1000}{\pi r^2}$$

Substituting this into the surface area formula:

$$\begin{aligned} A &= 2\pi r^2 + 2\pi r \frac{1000}{\pi r^2} \\ &= 2\pi r^2 + \frac{2000}{r} \end{aligned}$$

The area can be rewritten as

$$A = 2\pi r^2 + 2000r^{-1}$$

To determine the minimum surface area, we take its derivative and set it equal to zero:

$$\begin{aligned} \frac{dA}{dr} &= 4\pi r - 2000r^{-2} \\ 0 &= 4\pi r - \frac{2000}{r^2} \end{aligned}$$

After rearranging above equation, we can have the radius:

$$\begin{aligned} 4\pi r &= \frac{2000}{r^2} \\ 4\pi r^3 &= 2000 \\ r^3 &= \frac{2000}{4\pi} \\ &= \frac{500}{\pi} \end{aligned}$$

Finally, the value of r for minimizing surface area is as follows:

$$r = \sqrt[3]{\frac{500}{\pi}} \approx 5.42 \text{ cm}$$

Now, using $h = \frac{1000}{\pi r^2}$, we can determine the height of the oil can.

$$h = \frac{1000}{\pi \cdot (5.42)^2} \approx 10.84 \text{ cm}$$

Through the calculation with differentiation and the concept of calculus, Adam determines the dimensions to minimize material. The optimal dimensions for the one-liter oil can are:

- Radius is about 5.42 cm
- Height is about 10.84 cm

This scenario introduces a practical way to understand one of the key applications of calculus — optimization.

In real-world, manufacturers constantly seek ways to reduce costs while enhancing efficiency. This involves balancing two factors: the volume (which must remain fixed at 1 liter) and the surface area (which influences the amount of material used). This example not only enhances your understanding of differentiation but also illustrates how calculus is essential in design, engineering, and cost-saving decisions in everyday life.

Through these three different scenarios, we've explored how calculus is deeply connected to everyday decisions and challenges. Calculus helps us make smart decisions in business, understand how variables change over time, track motion and speed, and minimize materials in engineering and manufacturing.

These real-world examples show that calculus is not just abstract math but a powerful tool for problem-solving. Therefore, by learning calculus, we gain a new perspective on how the world around us operates mathematically.

9.4 Checkpoint 9

1. Revisit the example of Adam's Iced Coffee Stand. Previously, each \$1 increase in price reduced sales by 50 cups. Now, suppose that instead, sales decrease by 60 cups for every \$1 increase in price. How does this change affect the revenue function? Does the new price stay the same, or does it shift? Calculate the new price and compare it to the original case.

2. At a particular instant, Adam is biking north toward an intersection from 0.2 miles away at a speed of 5 mph. His friend Hannah is skateboarding east, already 0.4 miles past the intersection. A GPS tracker indicates that the distance between them is increasing at a rate of 3 mph. What is Hannah's speed at this moment?

3. In the earlier example, Adam designed a cylindrical oil can with a volume of 1 litre and was requested to minimize the amount of material used. The can must hold 2 liters instead of 1. How does this change affect the dimensions that minimize surface area? Calculate the new radius and height, and compare them to the 1-litre case. What do you notice about how doubling the volume influences the can's proportions?

9.5 Checkpoint 9 Solution

1. We are told that sales decrease by 60 cups for every \$1 increase in price. This gives us the demand function:

$$f(p) = 500 - 60p, \quad \text{with} \quad f'(p) = -60$$

The revenue function is the product of price and demand:

$$R(p) = p \cdot f(p) = p(500 - 60p)$$

and expanding:

$$R(p) = 500p - 60p^2$$

To maximize revenue, we take the derivative of $R(p)$:

$$R'(p) = 500 - 120p$$

Setting $R'(p) = 0$ to find the maximum revenue value:

$$\begin{aligned} 500 - 120p &= 0 \\ 120p &= 500 \\ p &= \frac{500}{120} \approx 4.17 \end{aligned}$$

The price that maximizes revenue is approximately \$4.17. As sales decrease by 60 cups for every \$1 increase in price, the revenue function adjusts accordingly. The new price is \$4.17 compared to the original price of \$5.00. Therefore, the price is lowered to maximize revenue when sales decline more sharply with each price increase.

2. Let x be Hannah's distance east of the intersection and y be Adam's distance south of the intersection. The distance between them is given by:

$$z = \sqrt{x^2 + y^2}.$$

At the given instant:

$$x = 0.4, \quad y = 0.2, \quad \frac{dy}{dt} = -5 \text{ mph}, \quad \frac{dz}{dt} = 3 \text{ mph}$$

We can derive Hannah's speed, $\frac{dx}{dt}$ after differentiating $z^2 = x^2 + y^2$ with respect to t:

$$2z\frac{dz}{dt} = 2x\frac{dx}{dt} + 2y\frac{dy}{dt}$$

and simplify as:

$$z\frac{dz}{dt} = x\frac{dx}{dt} + y\frac{dy}{dt}$$

First, calculate z at the instant:

$$z = \sqrt{x^2 + y^2} = \sqrt{0.4^2 + 0.2^2} = \sqrt{0.20} = 0.447$$

Substitute known values:

$$(0.447)(3) = (0.4)\frac{dx}{dt} + (0.2)(-5).$$

$$1.34 = 0.4\frac{dx}{dt} - 1$$

$$0.4\frac{dx}{dt} = 2.34$$

$$\frac{dx}{dt} = \frac{2.34}{0.4} \approx 5.85$$

Therefore, Hannah's speed is approximately 5.85 mph at that moment.

3. The oil can is a right circular cylinder with radius r (cm) and height h (cm). So, the surface area is expressed as:

$$A = 2\pi r^2 + 2\pi rh$$

The volume is 2 L as 2000 cm^3 and the height will be:

$$V = \pi r^2 h = 2000 \quad \Longrightarrow \quad h = \frac{2000}{\pi r^2}$$

After substituting h into the surface area, we obtain the surface area function with r as follows:

$$A(r) = 2\pi r^2 + 2\pi r\left(\frac{2000}{\pi r^2}\right) = 2\pi r^2 + \frac{4000}{r}$$

Differentiate with respect to r:

$$\frac{dA}{dr} = 4\pi r - 4000r^{-2}$$

Set $\frac{dA}{dr} = 0$ and solve:

$$\begin{aligned} 4\pi r - \frac{4000}{r^2} &= 0 \\ 4\pi r^3 &= 4000 \\ r^3 &= \frac{4000}{4\pi} \\ &= \frac{1000}{\pi} \end{aligned}$$

So, the radius is:

$$r = \sqrt[3]{\frac{1000}{\pi}} \approx 6.84 \text{ cm}$$

and the height is:

$$h = \frac{2000}{\pi r^2} \approx \frac{2000}{\pi (6.84)^2} \approx 13.68 \text{ cm}$$

The optimal radius for one-liter can was $r = \sqrt[3]{\frac{500}{\pi}} \approx 5.42$ cm and the height was about 10.84 cm. When the volume is doubled, the radius is $r = \sqrt[3]{\frac{1000}{\pi}} \approx 6.84$ cm and the height grows to 13.68 cm. Consequently, when the volume doubles, both the radius and height increase by $\sqrt[3]{2} \approx 1.26$ times.

Closing Thoughts

Have you had a chance to become familiar with derivatives in calculus? If you have a close friend whom you think highly of, you may want to introduce them to other people. Additionally, if others have unfairly judged said friend, you might feel frustrated and want to help others see their true qualities.

I wrote this book to share the essence of differentiation and calculus with as many people as possible. Many people react negatively to calculus due to improper learning methods. However, if you were able to let go of your predetermined notions about calculus through this book and encounter new aspects of calculus, that makes writing this book worth it.

The purpose of this book is to help readers understand the concept of differentiation and reimagine calculus through relatable examples from our daily lives. The concepts of derivatives in calculus that are introduced in this book are fundamental and essential principles that you must know before diving into the subject of calculus. If you have gained a good understanding of derivatives through this book, you are now ready to start learning calculus.

Summary

Throughout this book, we have explored how calculus provides a powerful way to describe and analyze change. The derivative arises from the fundamental need to express how quantities change mathematically. **Differentiation**, the process of finding a derivative function as the language of change, allows us to determine how a function behaves at any given point.

We distinguished between two key ideas: the **average rate of change** and the **instantaneous rate of change**. The average rate of change measures the slope between two points on a curve, expressed as:

$$\frac{f(b) - f(a)}{b - a}$$

However, this formula only describes an overall trend between two points. To understand the precise behavior of a function at a single point, we need the instantaneous rate of change — the slope of the curve at exactly one location. This leads to the fundamental definition of the derivative:

$$f'(a) = \lim_{x \to a} \frac{f(x) - f(a)}{x - a}$$

This formula connects two key ideas: the **limit**, which describes values approaching infinitely close, and the average rate of change between two points. Taken together, they show that the instantaneous rate of change at a point can be seen as the average rate of change as the two points become infinitely close.

Unlike the average rate of change, which requires two points, the instantaneous rate of change provides a precise measurement of how a function behaves at a single point. This makes it a powerful tool for accurately describing motion and change, particularly in fields such as science, engineering, and business. At the heart of this idea is bring the derivative, which mathematically expresses how quantities change at a specific instant.

Differentiation, the process of finding a derivative, gives us a systematic way to translate dynamic behavior into a mathematical framework — it is the **language of change**, allowing us to quantify and predict variations in real-world phenomena.

However, differentiation is only possible for continuous functions. Before we can determine the rate of change at a specific point, we must first ensure that the function behaves predictably at that point. In other words, motion cannot be analyzed at points where a function is discontinuous. The concepts of **continuity** and **discontinuity** can be formally defined as follows for a given point a.

$$\lim_{x \to a} \neq f(a) : f(x) \text{ is discontinuous at the point } a.$$

$$\lim_{x \to a} = f(a) : f(x) \text{ is continuous at the point } a.$$

The derivative formula can also be written in an equivalent form by letting $x = a + h$. This gives:

$$f'(a) = \lim_{h \to 0} \frac{f(a+h) - f(a)}{h}$$

Here, the difference between $f'(a)$ and $f'(x)$ is important: $f'(a)$ represents the slope at a specific point a while $f'(x)$ is a derivative function that allows us to find the slope at any point on the curve. The **derivative function** $f'(x)$ is **a function that explains the behavior of the instantaneous rate of change** for the entire function $f(x)$.

$$f'(x) = \lim_{h \to 0} \frac{f(x+h) - f(x)}{h}$$

By using the derivative $f'(x)$, we can uncover insights into the behavior or movement of the function. For example, if $f'(x) < 0$, the function $f(x)$ is decreasing. Conversely, if $f'(x) > 0$, the function $f(x)$ is increasing. Additionally, if $f'(x) = 0$, then $f(x)$ has a maximum or minimum. This relationship between the derivative and the function's behavior is a fundamental concept in calculus, providing a powerful tool for analyzing and understanding functions.

The derivative function $f'(x)$ may be expressed with the symbols shown below depending on the situation.

$$f'(x) = y' = \frac{dy}{dx} = \frac{d}{dx}[f(x)]$$

There are various differentiation rules, including the **power rule, sum and difference rules, product and quotient rules**, and the **chain rule.**

These rules can be summarized as follows. The derivative of a constant function, $y = c$, is always zero at any point since the constant function represents a horizontal line without any slope change throughout the function.

$$\frac{d}{dx}[c] = 0,$$

The derivative for functions with exponents has a pattern in which the degree of the exponent of x is reduced by one during differentiation.

$$\frac{d}{dx}[x^n] = nx^{n-1}$$

The derivative of the sum of two functions is equal to the sum of their individual derivatives. Similarly, the derivative of the difference of two functions is equal to the difference of their individual derivatives.

$$(f+g)' = f' + g' \text{ and } (f-g)' = f' - g'$$

However, the distribution property of the derivative does not apply to the multiplication and division of functions.

$$[f \cdot g]' \neq f' \cdot g'$$

$$\left[\frac{f}{g}\right]' \neq \frac{f'}{g'}$$

When two functions are connected to multiplication or division, the result of derivative functions can be obtained with the product or quotient rule shown below.

$$(f \cdot g)' = fg' + gf'$$

$$\left[\frac{f}{g}\right]' = \frac{f' \cdot g - f \cdot g'}{g^2}$$

When two functions are combined as the composition function, $f(g)$, the chain rule requires taking the derivative of the outer function and multiplying it by the derivative of the inner function.

$$[f(g)]' = f'(g) \cdot g'$$

$$[f(u)]' = f'(u) \cdot u'$$

We have seen that derivatives are more than just an abstract mathematical concept — they are a powerful tool for solving real-world problems. Real-world examples show that calculus surrounds us all the time, even when we don't notice it. Mathematical ideas or concepts are deeply rooted in our daily routines, helping us understand and connect with the world in meaningful ways.

Author's Note on References

The inspiration for writing this book partly comes from How to Lie with Statistics by Darrell Huff, which shows statistical concepts through everyday examples. Some explanations of function properties are based on Foundations and Pre-Calculus Mathematics 10 by Davis and Garry. The presentation of calculus ideas — including key formulas and definitions related to derivatives — follows standard treatments such as Calculus: Early Transcendentals by Howard Anton, Irl Bivens and Stephen Davis.

Notes

Some of the approaches and examples in this book are similar or inspired by *Foundations and Pre-Calculus Mathematics 10* by Davis and Garry, and *Calculus: Early Transcendentals*, 9th ed., by Anton et al. In particular:

14. Table illustrating slope (see Davis, p. 301).
16. Function machine diagram (see Davis, p. 268).
17. Explanations of functions and related examples (see Davis, pp. 265-269).
33. Graphs showing x approaching 2 from the left or right side (see Anton, p. 70).
34. Table illustrating approaching 2 from the left or right side (see Anton, p. 70).
40. Graph showing that the secant line slope approaches the tangent line slope (see Anton, p.132)
56. Graph showing a removable discontinuity (see Anton, p. 110).
86. Graph showing that the secant and tangent line when x is h distance away from a (see Anton, p.133)

93. Common Symbols for Derivatives (see Anton, p. 150).

118. Explanation of a product rule example (see Anton, pp. 163-164).

141. Scenario 1: Adam's Iced Coffee Stand (see Anton, p. 168).

146. Scenario 2: A Related Rates Bicycle and Skateboard (see Anton, p. 206)

149. Scenario 3: Minimizing Material Costs (see Anton, p. 279).

Bibliography

Howard Anton, Irl Bivens and Stephen Davis. Calculus: Early Transcendentals. 9th ed., John Wiley & Sons, 2009.

Davis, Garry. Foundations and Pre-Calculus Mathematics 10. Pearson Education Canada, 2010.

Huff, Darrell. How to Lie with Statistics. W. W. Norton & Company, 1954.